INTERNATIONAL MINING FORUM 2015

23–27 FEBRUARY 2015
CRACOW, POLAND

Vertical and Decline Shaft Sinking – Good Practices in Technique and Technology

International Mining Forum 2015

Edited by

Jerzy Kicki
AGH University of Science and Technology, Cracow, Poland
Polish Academy of Sciences, Mineral and Energy Economy Research Institute, Cracow, Poland

Eugeniusz J. Sobczyk
Polish Academy of Sciences, Mineral and Energy Economy Research Institute, Cracow, Poland

Paweł Kamiński
AGH University of Science and Technology, Cracow, Poland

CRC Press
Taylor & Francis Group
Boca Raton London New York

CRC Press is an imprint of the
Taylor & Francis Group, an **informa** business

A BALKEMA BOOK

Published by: CRC Press/Balkema
P.O. Box 11320, 2301 EH Leiden, The Netherlands
e-mail: Pub.NL@taylorandfrancis.com
www.crcpress.com – www.taylorandfrancis.com

First issued in paperback 2020

© 2015 by Taylor & Francis Group, LLC
CRC Press/Balkema is an imprint of the Taylor & Francis Group, an informa business

No claim to original U.S. Government works

ISBN 13: 978-0-367-73845-7 (pbk)
ISBN 13: 978-1-138-02820-3 (hbk)

Visit the Taylor & Francis Web site at
http://www.taylorandfrancis.com

and the CRC Press Web site at
http://www.crcpress.com

Typeset by Krzysztof Stachurski

International Mining Forum 2015, Kicki et. al. (eds) © 2015 Taylor & Francis Group, London, UK. ISBN 978-1-138-02820-3

Table of Contents

International Mining Forum 2015, Kicki et. al. (eds) © 2015 Taylor & Francis Group, London, UK. ISBN 978-1-138-02820-3

Preface

The International Mining Forum is a meeting of scientists and professionals who, together with the organizers, establish ambitious aims to confront ideas and experiences, evaluate implemented solutions, and discuss new ideas that might change the image of the mining industry.

The gained technical and technological experience allowed for development of design methods and improvement of technology. In particular it was related with shaft lining design methods. The present research try to explain these problems to designers, including demonstration that design of underground excavations lining (particularly shaft lining) should be considered as a process of modifications of procedures accompanying gathering knowledge about properties of the rock mass, including lining material.

The International Mining Forum is an event accompanying the School of Underground Mining, organized since 1992 under the auspices of AGH – University of Science and Technology and the Institute of Mineral Economy and Energy of the Polish Academy of Sciences of Cracow.

The topics presented during the session will be related to the implementation and operation of shaft facilities. Each of the interesting presentations will be in the range of discussion topics, such as geological and hydrogeological conditions and the technical operation of shafts, overview of construction technology, basic issues related to the design process of sinking and shafts lining, directions aimed at modernizing the technologies of sinking.

The organizers would like to express their gratitude to all persons, companies and institutions, who contributed to bringing the Forum into being.

We hope that the Forum will contribute to the exchange of interesting experiences and establishing new acquaintances and friendships.

Jerzy Kicki
Chairman of the IMF 2015 Organizing Committee

International Mining Forum 2015, Kicki et. al. (eds) © 2015 Taylor & Francis Group, London, UK. ISBN 978-1-138-02820-3

Organization

Organizing Committee:
Jerzy Kicki (Chairman)
Jacek Jarosz (Vice-Chairman)
Eugeniusz J. Sobczyk
Artur Dyczko
Dominik Galica
Rafał Polak
Michał Kopacz
Paweł Kamiński

Advisory Group:
Volodymyr Bondarenko, professor (National Mining University of Dnipropetrovsk, Ukraine)
Alfonso Carvajal (Universidad de La Serena, Chile)
Piotr Czaja, professor (AGH University of Science and Technology, Poland)
Jozef Dubiński, professor (Central Mining Institute, Poland)
Roman Dyczkowsky professor (National Mining University of Dnipropetrovsk, Ukraine)
Jaroslav Dvoracek, professor (Technical University VSB, Czech Republic)
Eugeniusz Mokrzycki, professor (Polish Academy of Sciences, Mineral and Energy
    Economy Research Institute, Cracow, Poland)
Jacek Paraszczak, professor (Laval University, Quebec City, Canada)
Anton Sroka, professor (Polish Academy of Sciences, The Strata Mechanics Research Institute,
    Cracow, Poland)
Mladen Stjepanowic, professor (University of Belgrade, Republic of Serbia)
Antoni Tajduś, professor (AGH University of Science and Technology, Poland)
Yuan Shujie (Anhui University of Science and Technology, Huainan, Anhui, China)

International Mining Forum 2015, Kicki et. al. (eds) © 2015 Taylor & Francis Group, London, UK. ISBN 978-1-138-02820-3

Future Trends in Shaft Development

E. Neye, W. Burger, A. Frey,
D. Petrow-Ganew, Benjamin Künstle
Herrenknecht AG, Germany

ABSTRACT: Mining companies have to break new ground to further modern, efficient and safe future mining projects. Herrenknecht supports companies on this path with innovative machine technology. For mining companies accessing deep mineral deposits or developing underground mine infrastructure the aim today has to be to create safe and attractive work places for personnel while enhancing the shaft excavation rates significantly. Mechanized shaft development is the way to go.

The application of mechanized shaft sinking machinery will lead to an increase of the overall performance due to simultaneous succession of work steps and thus the net present value of the project. Herrenknecht AG has developed in close contact to the mining industry shaft excavation applications for different shaft and project requirements to match the industry's needs.

BACKGROUND

In the past decades the pressure on the industry to develop mineral deposits in adverse geological or climatic condition or to extend existing mining projects throughout the world was and is increasing. The need to build new shafts or to modify the infrastructure of an existing underground mine has been driving the industry desire for more competitive technologies in shaft sinking with the focus on safety and efficiency.

Novel developments must push and move beyond the current performance limits of conventional shaft sinking technologies. Each development needs to have the objective of a significant technical innovation, improved performance and cost reduction, while enhancing the shaft crew's safety.

Mechanical excavation methods can significantly improve excavation performance and labor safety compared to drill and blast operations. With its developments in shaft sinking machines, boxhole boring units and raiseboring rigs Herrenknecht meets the industry's demand for fast and efficient shaft development technologies by incorporating highest safety engineering.

MECHANISED BLIND SINK METHOD

Shaft sinking is the critical path of a mining project; it has been identified as one of the highest risks undertaken in the mining industry. Until today the majority of deep shaft excavation is done using the traditional drill and blast method. Mechanical excavation method are a step-change in excavation performance, as productions rates are considerably higher, and labor safety enhanced compared to drill and blast operations. Saving time on shaft sinking activities and thus achieving a quicker access which in turn enables a quicker build-up of the production profile, is the real benefit, as it increases the Net Present Value (NPV) of a project.

Taking a look at the process of shaft sinking, blind sink methods consist usually of 4 phases:
(1) Establishment of the shaft collar, pre-sink and shaft sinking infrastructure.
(2) Commissioning and installation of shaft sinking equipment.
(3) Routine phase of shaft sinking cycle for crew, improvement of processes.
(4) Stripping of shaft sinking equipment from shaft bottom.

The general sequence of those four phases will not change when comparing drill and blast operations with mechanical shaft sinking activities.

The more basic equipment of drill and blast operations leads to a shorter time to operation then with the more complex machinery of the mechanized shaft sinking.

Mechanized shaft sinking will show its full effect in phase 3, during the actual shaft sinking phase. When looking at the advance cycles for both operations, the application benefit of mechanized shaft sinking becomes evident.

When applying mechanized shaft sinking technologies work steps of excavation, material pick up and shaft lining can be performed in parallel where with conventional drill and blast operations each step has to wait for its predecessor to be finalized.

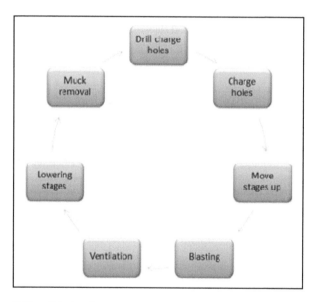

Drill and blast cycle

The advantages of cutting methodologies in blind shaft sinking have to be an increased safety by removing employees from the danger areas, increase shaft sinking rates by utilizing cutting edge technology and being cost effective by increasing the productivity of the project.

A first step to meet those criteria was the development of the so called VSM system in the 1990th. The system is designed to work in water-bearing soils and soft rock. This system is based on roadheader excavation and operates in submerged conditions. For the muck transportation from the shaft bottom a slurry circuit is installed. Due to restrictions given by the slurry pressure at the shaft bottom the maximum depth of the shafts are limited at approx. 100 m.

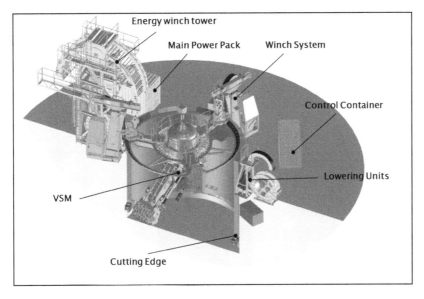

VSM Technology

The system is used successfully in civil construction to build shafts for sewage tunnels or ventilation shafts for subway tunnels as an alternative solution for the slurry wall method.

In the next approach Herrenknecht started to tackle the requirements for deep blind sink methods.

Even though developed for very different application areas and project requirements, the three basic functions excavation, rock support and the provision of installation decks for supply infrastructures are common. The machines consist of three system areas. System area 1 is a no-go zone for personnel while the machines are in advance mode, only during maintenance time's personnel is allowed to enter the excavation chamber. In system area 2 all preliminary shaft wall support works are carried out. The machines can be developed to support rock bolting, shotcrete or concreting activities. In system area 3 all permanent work places and the machines infrastructural components are located.

The Shaft Boring Machine (SBM) is a development for mechanized excavation of deep vertical blind shafts in hard rock conditions in cooperation with Rio Tinto. The semi-full-face sequential excavation process is based on the use of a rotating cutting wheel excavating the full shaft diameter in a two stage process for one complete stroke.

The excavation process is divided into two steps. With the first step a trench is cut with the cutting wheel rotating around its horizontal axis and being pushed downward by cylinders. With the second step of excavation the entire bench area is excavated by slewing the rotating cutting wheel 180° around the shaft's vertical axis.

The cutting wheel circumference and periphery of both sides is equipped with appropriate cutting tools and muck buckets to excavate the rock and remove the cuttings while rotating. Excavation and muck removal is a continuous process. The cuttings are guided along internal muck channels and discharged by gravity onto a center arranged secondary conveying system.

All reaction forces of the excavation process are transferred into the shaft walls by grippers. During the gripper reset operation after each excavation cycle the machine can be adjusted along its vertical axis for alignment control.

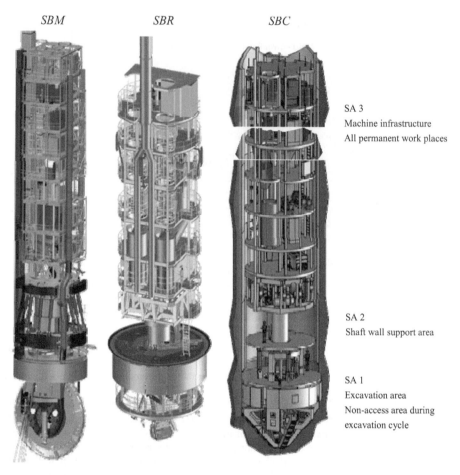

SBM SBR SBC

SA 3
Machine infrastructure
All permanent work places

SA 2
Shaft wall support area

SA 1
Excavation area
Non-access area during
excavation cycle

Machine set-up

Subsequently the so called Shaft Boring Roadhead (SBR) was developed to meet the industry's requirement for applications in dry soft to medium hard rock with variable diameters. The SBR is a shaft sinking machine suspended by ropes connected to shaft winches.

The cutting is done using a rotating cutting drum equipped with cutter picks. The hydraulic or electric driven cutter drum is installed at the front end of a telescopic boom. The articulated telescopic boom can rotate +/- 200° around the shaft axis and so cover the complete bench area. One excavation cycle of 1.0 m depth consists of five sub-cycles of each 200 mm steps using the cutter boom telescopic function. After a full excavation cycle is finished the entire SBR is reset and lowered down by 1.0 m and a new cycle can start.

The excavation sequence is fully automated and cuts the material without disturbing the surrounding ground and with almost zero over-break. The automated cutting procedure reduces operator errors and allows for a constant cutting process. The operator controls the cutting sequence from an operator cabin located in the S3 area away from the shaft bench. All systems are monitored and controlled by the operator via computer screens including a data recording system. The collected data can be transferred to surface to be analyzed or monitored off site.

For cutter pick maintenance the SA1 area is accessible in a standstill mode. If required drill rigs for probing or pre excavation grouting can be installed in the SA1 section. During a normal excavation cycle of 1.0 m, rock bolts and shotcrete can be applied to the shaft wall from the rock support platform in the system area 2 (SA2) simultaneous with the excavation process.

The **Shaft Boring Cutterhead** (SBC) is a mechanical shaft sinking unit for the excavation of deep blind shafts in hard rock conditions with an inner diameter of up to 8.6 m. The SBC is suspended and moved by the shaft ropes.

The SBC utilizes a conical shaped cutterhead equipped with disc cutters. The cutting sequences will be highly automated. Within the excavation chamber a certain water level will be maintained by a water circuit so that the cutterhead works submerged.

Mucking of cut material will be carried out by the shaft hoisting system for coarse material and via hydraulic conveyance for fine material.

The SBC will be equipped with a rock bolting unit and a unit to install the shaft lining. Unless there are extremely poor ground conditions, rock support activities and the excavation process can be carried out simultaneously.

Each machine component has been designed to reach an instantaneous cutting rate of 6 meters per day.

During the regular excavation process, personnel working on the SBC are not exposed to the unsupported shaft wall and hence falling rock, they are also not exposed to silica dust. No explosives are required or used during a regular SBC advance operation.

Regular work areas (work decks) and the excavation chamber are separated by a bulkhead, the so called gripping unit. The excavation chamber is a "no go zone" during regular excavation.

NEXT DEVELOPMENT STEP – MODULAR CONCEPT

Taking a look at the machines suspended by shaft ropes it becomes obvious that the various machine options can be attached to the basic machine components. To facilitate and reduce the machine design phase the next development step has to be to create a modular concept for the machines suspended by shaft ropes; a toolbox where to the customer's project needs the machine can be assembled by existing machine modules.

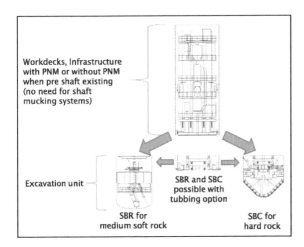

Modular machine set-up

Similar to all machine types will be the setup of three system areas of excavation chamber, primary rock support and permanent workplaces and machine infrastructure. On the project requirements will depend and vary the topics of:

- Muck transportation system.
- Type of primary rock support.
- Type of excavation unit.

HERRENKNECHT RAISE BORING RIGS FOR SHAFTS UP TO 2000 M DEPTH

With growing demand for raw materials, mines are applying increasing effort and expense to open up deep and difficult to access resources. With this in mind, the new Herrenknecht product portfolio of Raise Boring Rigs allows the optimal drilling of shafts with larger diameters at depths of up to 2000 meters. In the new and further development of Raise Boring Rigs, Herrenknecht draws on proven technologies and components as well as on extensive know-how in the area of mechanized tunnelling technology.

The Raise Boring Rigs from Herrenknecht have proven themselves worldwide in a variety of projects since the first use of a prototype in 2010 for the Vianden hydropower plant in Luxembourg [1]. They have demonstrated their efficiency both in drilling of production and ventilation shafts in mines (copper, coal, tin) as well as in infrastructure projects (pressure shafts for hydropower plants, ventilation shafts for road tunnels) (Table 1: all projects completed to date). Today Herrenknecht offers a product portfolio of four RBR types.

MINE SITE/PLACE	COUNTRY	DIAMETER [m]	SHAFT DEPTH [m]	MARKET
Vianden	Luxemburg	5,46	280	Hydropower
Bolzano	Italy	4,76	280	Civil
Venda Nova	Portugal	5,46	350	Hydropower
Nant de Drance, 1	Switzerland	2,4	420	Hydropower
Chuciquamata	Chile	3,5	350	Mining
Bergamo	Italy	3,68	200	Mining
River View Coal Mina	USA	4,3	90	Mining
Huanuni Mine	Bolivia	3,06	400	Mining
Nant de Drance, 2	Switzerland	2,4	420	Hydropower

RAISE BORING RIG (RBR)

The Raise Boring Rig (RBR) developed by Herrenknecht is designed for the construction of shafts in hard rock down to depths of 2,000 meters. Reaming shafts with RBRs is safer, less labor intensive and more cost effective than conventional shaft sinking, which was previously the only possible method beyond 1200 meters.

With its compact design, the RBR offers high flexibility even in confined spaces and is therefore suitable for a variety of applications in the mining industry. It creates shafts for the transport of muck or ore, haulage shafts, pressure shafts for hydropower plants and supply shafts for energy, water and air. It has proven itself in use generally with its modular design and its powerful and highly efficient center-free *drive*.

METHOD OF OPERATION

Initially the rig is installed above the collaring point with the crawler unit or a crane. It drills the pilot hole downwards vertically or at an angle of up to 45 degrees with the drill bit. Depending on the drilling depth, further drill rods are installed progressively, until the target in an already existing tunnel or cavern is reached. The pilot hole drill bit is then removed in the cavern and the reaming head or reamer is connected to the drill string. After completion of the assembly work, the rig pulls the reaming head equipped with cutters upwards against the face. The rotation of the rig is transferred via the drill string to the cutterhead, which in combination with the contact pressure of the cutters crush the rock. The material falls down and can be easily removed. In this way, the entire shaft is reamed upwards to the required diameter.

PRODUCT PORTFOLIO OF HERRENKNECHT RAISE BORING RIGS

The product portfolio of Raise Boring Rigs from Herrenknecht includes the types RBR300VF, RBR400VF, RBR600VF and RBR900VF (Fig. 1). With power ratings between 300 kW and 800 kW (402 hp – 1,072 hp) and thrust forces between 458 t (1,009,717 lbf) and 2,243 t (4,944,969 lbf), they cover a wide range of applications.

Drilling shafts up to 2,000 meters with large diameters of up to eight meters requires rigs with high torque and high thrust forces. This need is targeted particularly by the Raise Boring Rig RBR900VF with torque of 900 kNm (663,805 ft-lbf) and thrust of 2,243 t (4,945,797 lbf), developed by Herrenknecht in cooperation with Australian mining contractor Macmahon. The RBR-900VF is the most powerful Raise Boring Rig currently on the market.

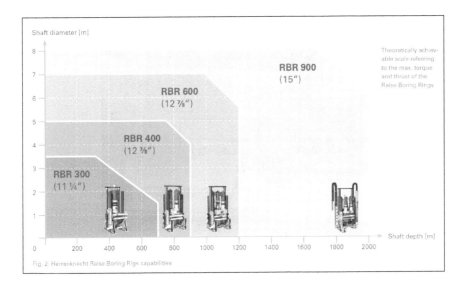

Fig. 2: Herrenknecht Raise Boring Rigs capabilities

In addition to its high performance the RBR900VF is distinguished by the automated drill pipe feeder developed by Herrenknecht. Compared with manual handling, the remote controlled system ensures both efficient workflows as well as significantly greater work safety for the personnel during installation and removal of the drill string.

Herrenknecht Raise Boring Rigs are generally used over several years and in different projects. Besides their high performance and operational safety they are distinguished by their reliability and long service life. Their design allows quick installation of spare parts and flexible adjustments of their performance: all rig types have a powerful and highly efficient center-free drive. Multiple identical motor gear units are arranged around the center of the drive, so that even if one of the motors fails, work can continue with reduced power without interruption. This drive system also allows additional equipping with more motors. For example, a standard RBR600VF can subsequently have an additional 4th motor retrofitted. The total drive power and the available torque are consequently increased so that the rig can be used for larger shafts and drill rod diameters.

Effectiveness and precision reduce wear and thus additionally prolong service life. The tried and tested frequency converter controlled drive concept used in Herrenknecht tunnel boring machines means energy consumption is lower. The electric motors achieve significantly higher efficiency compared to hydraulically powered machines. Variable speed and torque control also allows the precise transmission of power to the drill string. This increases the efficiency of the drilling operation without risking overloading of the individual drill pipes. With heavy drill rods in particular, the innovative active float concept designed by Herrenknecht can also reduce the load on threaded connections when screwing and unscrewing. Lifting the floating box by means of a hydraulic cylinder reduces wear and extends the life of the drill rods.

All Herrenknecht rigs are characterized by a compact and modular design. Transport of the rigs can be by road with trucks without special transports and by ship in standard sea containers.

PROJECT DEPLOYMENTS OF HERRENKNECHT RBRS
NANT DE DRANCE, SWITZERLAND

Marti Contractors Ltd used two RBR600VF rigs to bore two vertical pressure shafts for the new Nant de Drance pumped storage power plant (Canton of Valais, Switzerland). Between August 2012 and November 2013 the two rigs drilled two shafts with a length of 420 meters and a diameter of 2.4 meters through rock with strengths of up to 150 MPa.

The top daily performances were 62 meters (pilot hole) and 50 meters (reaming). The reaming of the second shaft was successfully completed after nine days and an average drilling performance of 46 meters a day.

To improve work safety for the operating personnel, the second Nant de Drance rig Marti ordered was equipped with an automatic wrench unit from Herrenknecht. It replaces the conventional manual attachment method of screwing and unscrewing the drill rods that the personnel otherwise have to do directly on the rig. In addition, the automatic wrench unit speeds up installation and removal of the drill string. This optimization in the workflow has a positive effect especially when a complete drill string has to be removed and re-installed for servicing.

With the forklift equipment that can be attached to the crawler unit Herrenknecht implemented another innovation to meet customer requirements: the rig's crawler unit, which usually remains unused during the drilling process, was designed so that it can be used on site as a forklift. Thus a high degree of autonomy is achieved for the operation of the rig, because components and equipment as well as power packs can be transported independently and rapidly on site.

After completion of the RBR drilling in Nant de Drance a Herrenknecht Shaft Drilling Jumbo was used to enlarge the shafts to a diameter of eight meters. In addition, Marti Tunnelbau AG used a Herrenknecht tunnel boring machine (Gripper TBM, Ø 9,450 mm) in the Nant de Drance project to drive a 5.6 km long access tunnel to the machine cavern of the power plant.

SUMMARY AND OUTLOOK

The newly developed Raise Boring Rigs from Herrenknecht have proven themselves worldwide in numerous project deployments. Altogether, nine shafts with a total length of 2800 meters have been created with the Herrenknecht rigs in mining projects as well as in hydropower plant and road construction projects.

The Herrenknecht RBR product portfolio targets the growing need for deep shafts with large diameters and the development of project-specific, innovative technologies. On the basis of many years of experience as a market and technology leader in the area of mechanized tunnelling, Herrenknecht actively supports clients and projects with its great innovation capacity and its global service network as well as with short reaction times.

BOXHOLE BORING MACHINE TECHNOLOGY

The **Boxhole Boring Machine** (BBM) developed by Herrenknecht is designed for the excavation of slot holes and vertical or inclined small-diameter shafts. The BBM concept is based on the well-proven microtunnelling pipe jacking technology – an area in which Herrenknecht can look back on 25 years of experience in international projects.

The advantage of the developed BBM system is the blind boring concept which allows a wide range of application in the mining field. With this new technology the drilling operation has become independent from the development of an upper level. Also it could be used when it is financially more feasible than a raised hole. The main applications in the mining field are the draw bell slot holes for block/panel caving operations, ventilation shafts, ore passes, service shafts and tabular narrow veins exploitation.

Increased safety, high mobility, quick relocation, high performance and minimum space requirements were the key factors considered during the development of this machine to match the technical and functional requirements in an underground mine application.

The product portfolio of Boxhole Boring Machines from Herrenknecht includes the types BBM1100 and BBM1500. The names indicate the drilling diameter in millimeter. With power ratings between 160kW and 200kW, maximum torque of 135kNm and maximum thrust force of 2,500 kN, they cover a wide range of applications. In cooperation with Codelco, the largest copper producer in the world, a feasibility study has been carried out to build larger machines up to 2000 millimeters in diameter.

METHOD OF OPERATION

Two elemental components of the BBM supply the necessary movements for drilling: the jacking frame which provides the thrust force and the boring unit which provides the rotational motion. The power unit is the electro-hydraulic supply of the complete system and the hose and cable drum transmits all media and signals required for the boring operation. To ensure the BBMs fast transport to production areas in confined underground conditions, the BBM system is mounted on a compact transport unit. The BBM does not need preliminary preparations (concrete foundation). The only requirement would be a leveled ground of sufficient load capacity. Once the installation site has been reached, the unit is then deployed and set into the proper orientation. To stabilize the system and transfer the operational thrust and torque loads into the rock, the jacking frame is braced against the floor and the back by hydraulic feet and grippers. The thrust pipes are brought to the BBM and mounted by a fork lift or other suitable mobile equipment. The thrust plate can then be pushed against the pipe string, advance speed and thrust force is fully adjustable.

The cuttings are guided by gravity in a channel through the pipe string and finally into the skip located next to the jacking frame. Once a pipe is drilled, the next pipe is installed by a fork lift and the drilling process can be restarted. Once the machine has reached the final hole length the entire pipe string is pulled back in steps while the pipes are removed piece by piece from the jacking frame.

BBM1500 Field test in Grube Clara mine. Blackforest, Germany. March 2013

PROJECT REFERENCES

Mancala Ltd., a company specialized in design, engineering, construction, excavation and mining services, counts with a fleet of three BBM1100 boring successfully more than 90 slots and 1,500 meters in total up to now in different mines in Australia. The first one deployed at Cadia East mine, achieved very good advance rates of 2 m/hr including drilling and pipe installation. In Average, one hole with an average length of approx. 17 m could be drilled in 1.5 to 2 work shifts.

Additionally two BBM1500 are currently deployed in Chile at El Teniente mine, owned by Codelco. Two experienced Chilean contractors, Mas Errazuriz and Gardilcic, operate independently these machines, using them up to now mainly for ventilations shafts. Both machines together have drilled in total more than 20 slots and 600 meters since July 2013.

Mine site/place	Country	Diameter (mm)	Shaft depth (m)	Market
Cadia East	Australia	1100	up to 30	Mining
El Teniente	Chile	1500	up to 60	Mining
Frog Legs	Australia	1100	up to 30	Mining
Perilya	Australia	1100	up to 30	Mining
El Teniente	Chile	1500	up to 60	Mining
Sunrise Dam	Australia	1100	up to 30	Mining

REFERENCES

Worley Parsons TWP, Feb. 2014: Overview of Mining Studies and the Benefits of Quicker Access to the Ore Body.

P. Schmäh, B. Künstle, N. Handke, E. Berger: Glückauf, 143, 2007, Nr 4, Weiterentwicklung Und Perspektiven Mechanisierter Schachtteuftechnik.

E. Neye, W. Burger; Aims 2013: Rapid Blind Shaft Sinking.

W. Burger, F. Delabbio, C. Frenzel; SME Annual Meeting 2010: Accessing Deep Ore Bodies Using Mechanical Excavation Equipment.

W. Burger, M. Rauer; SME Annual Meeting 2012: Mechanized Shaft Sinking Method for Soft and Medium Strength Rock.

H. Hartman et al.: Society for Mining, Metallurgy and Exploration Inc. Littleton, Colorado 1992, SME Mining Handbook.

Take It to the Next Level. New Technology for Creating Slot Holes. M. Stöhr, B. Künstle & W. Burger, Herrenknecht AG, Germany.

Künstle B., Frey A. 2011: Vertical Shaft Construction at the Pump Storage Plant Vianden/Luxembourg. Tunnel 4/2011, 56–59.

International Mining Forum 2015, Kicki et. al. (eds) © 2015 Taylor & Francis Group, London, UK. ISBN 978-1-138-02820-3

Investment Costs Comparison
of Shafts and Declines

Branko Gluščević, Čedomir Beljić, Zoran Gligorić
Faculty of Mining and Geology, University of Belgrade, Serbia

ABSTRACT: In this paper we focus on some of the more important economic criteria that should be taken into account when planning the variants of the opening with shafts and/or declines systems, required for exploitation of the mineral deposit.

The complexity of the problems that are encountered and solved when mining is started should be taken into consideration since it is evident that projects are unavoidably of the multidimensional nature and frequently controversial. We think that multidimensional nature of the mining projects may be rather simple, and in a certain way that multidimensional nature may be distorted and presented as a single criterion.

Optimization of the financial return per rand of venture capital employed for the project as a whole may be our choice for the selection the parameters that are to be designed. This criterion may not always satisfy the investment objective in all cases (as, for instance, where risk limitation is important), however it may serve to place the relative economic advantages and disadvantages of specific design features, into correct perspective with respect to the overall economics of the project under the prevailing economic ambient at the time of the investigation, although they may be sometimes in opposition.

KEYWORDS: Underground mining, shafts and declines, investment costs, multi-criterion optimization

1. INTRODUCTION

Trends in world production of mineral raw materials have the tendency of moving in the direction of mass production and large capacities in big deposits which in most cases are very poor. Such orientation cannot completely eliminate certain role and significance of small deposits. Exploitation of small deposits in the world is known as "small-scale mining" or "restricted range mining" and it still exists due to several reasons. Main reasons that are in favour of small deposits exploitation are linked on one hand to the genesis itself of these deposits, and on the other hand to exceptionally favourable technical, organizing and financial possibilities in the course of their exploitation. Small deposits require relatively short period of time needed for prospecting and rather small investments in the course of prospecting and exploitation. Possibility of engaging local labour, as well as relatively fast way to secure certain quantities of high quality raw materials have influence on the fact that many countries in the world successfully keep up this kind of exploitation.

Production from small deposits provides around 10% of total world production of various mineral raw materials; however there is still no strict international classification which defines the idea of "small deposits" and small mining capacities. These problems are mainly connected with the so called minimal reserves that in the economic sense make possible simple reproduction in concrete exploitation conditions. Minimal reserves are in the function of the kind of raw minerals, type of the deposit within the similar raw mineral, region, and country in which the exploitation is taking place.

2. FORMER EXPERIENCE IN SMALL-SCALE DEPOSITS EXPLOITATION

Mining of small capacities, disregarding the differences in defining the mentioned criteria is mostly distinguished by the following characteristics:
– relatively low degree of deposit prospecting and of the knowledge of mineral material quality,
– non-systematic exploitation due to lack of data on deposit,
– low capital investments due to significant participation of living labour,
– relatively low degree of utilization of the mineral material when exploiting and a high degree of dilution,
– unfavourable working conditions, especially in underground exploitation,
– inadequate social care and insufficient safety of workers,
– low tax revenues for the government, etc.

However it is also necessary to single out those characteristics of small capacities mining that are regarded as their advantages:
– short prospecting period and investing of relatively meagre financial means,
– possibility to build fast mining capacities and installations for mineral processing, with rather small investments and relatively small energy consumption,
– equipment is often light and simple, also exceptionally mobile which is a great advantage especially when exploiting silt deposits of gold, precious stones and semi-precious stones, etc.,
– possibility to employ local labour which effectively reduces unemployment in undeveloped regions,
– smaller quantities of needed deficient mineral material may be relatively quickly secured,
– mineral materials from such deposits are mostly of a high quality, and that is secured by way of various procedures of their preparation,
– these mines cause far less negative results on the environment.

All negative characteristics are subject to changes for the better by introducing more modern technology, and that mostly depends on economic factors, that is, the value of the mineral material.

3. OPENING OF SMALL-SCALE DEPOSITS

The initial phase of mining activities in the process of underground exploitation is called opening of the deposit. It is characterized by making a system of mine rooms with basic objective to secure communication between ore deposit and the surface of the terrain.

Depending on the conditions in which the exploitation is taking place during the phase of opening development we can recognize: different types of room openings, organizing forms, as well as characteristic ways of using the same. Here will be presented some of the usual ones: classification of the opening system, influential factors, as well as the special objectives that are realized by this exploitation phase. Generally speaking, the fact that exploitation conditions are becoming drastically complex may serve as the basic characteristic of the opening, and that can be noticed in several aspects such as: utilization of the existing ore deposits, content reduction of useful components in ore, changeable market conditions when talking about exploitation of metallic mineral raw material.

The consequence of the mentioned is difficulties in business deals as well as aggravation of total economic effects of exploitation. It can be said that underground exploitation is taking place at greater and greater depths, especially in localities where there are, according to various criteria, rich ore deposits. On the other hand there occurs increase in the quantity of mass in the existing deposits, and that has been made possible by application of new technologies.

There are more and more mine openings with small-scale and medium-scale annual production capacities. Thus established flexible production units have the possibilities to deal in their own way with problems caused by the changeable business conditions. Specific organization plans starting from those that make possible production of some ten million tons of ore annually, and then mining at the depths greater than 2,000 m to small-scale mines with few employed, practically at the level of the family manufacture, all these mean application of various technical and technological solutions. So the manner in which the mines are opened as well as preparations give a wide scope of solutions. What solutions are to be developed and used depend on a number of influential factors.

Mining practice knows general classifications of the opening systems. However, as much as these classifications are useful and applicable, the fact is that every mine or potential deposit are specific to a significant degree and the schematized approach for the solving of problems cannot always be applied. It seems necessary to point out that in literature there are very different classifications by which the system is described, but depending on the criteria taken into consideration. Also, very often the classifications given by various authors are in fact similar, or differ in details only. Criteria and classifications that have been pointed out represent specific selection and systematization and are based on a large number of data from literature.

3.1. Classification of the Opening System

Planning and designing of the opening phase means solving a number of problems which are:
- selection of the type and definition of the constructive characteristics of main and auxiliary rooms for opening;
- place for location and arrangement for rooms for opening;
- number of rooms for opening;
- number of basic horizons, i.e. mining levels;
- defining technological complex on the surface;
- safety measures;
- transport and hauling of ore, waste, labour, repro-material;
- energy supply;
- ventilation, drainage and pit servicing.

Pit opening may be complete or partial. Complete opening means fully opening the deposit which is being prospected, and that is in practice rarely done. This term even is causing a slight disagreement. Namely some authors think that complete opening is a theoretical category taking into consideration impossibility to completely define ore reserves in the course of prospecting phase.

Partial opening means opening of the parts of the deposit or of single mineral bodies. This means that later on, when considered appropriate, remaining reserves may be opened. This is frequently done in practice and is based on the approach that mining starts at the moment when balance reserves are defined, that is, when we have defined ore quantities that can be economically mined.

Potential reserves are eventually additionally prospected, categorized and opened parallel to the exploitation process. As one of the criteria for the beginning of exploitation of the parts of deposit literature gives that it is necessary to define reserves for 10–12 years of exploitation, that is, everything depends on the dynamics of achieving the appropriate production capacity.

For different mineral materials these terms are different. They are mostly defined on the basis of demands for economic business dealings.

In any case opening of the entire or parts of the deposit has to be economically profitable. Exploitation term has to be accepted, besides other things, as a factor in the function of economics.

The manner of ore deposit opening is defined by the type and layout of the main objects and rooms that are to be opened as well as the number of additional objects.

The term, main objects to be opened, in this case means main haulage and ventilation rooms as well as transport and ventilation rooms that are connected to the main haulage and ventilation, and have a major character.

Auxiliary rooms for openings are service shafts, crosscut drifts, filling stations of the shaft, sumps, pump stations, as well as the other rooms that have been designed to be used for a longer period of time.

According to the type of the room the manner of opening may be:
- by horizontal rooms,
- inclined rooms,
- vertical rooms, and
- by combined rooms.

Opening of ore deposits has to be in agreement with application of effective means for transport and mining, transport of people, pit servicing, drainage, ventilation etc. In other words, by opening system are secured functioning and exploitation of the mining mechanization.

Table 1. Classification of factors by nature and characteristics

Natural-geological factors	Technical technological factors	Infrastructural factors	Economic factors
location of deposit in Earth's crust	building speed, term for exploitation start up	expenses for construction of opening objects	transport situation
depth of occurrence	production capacity	expenses for equipping opening rooms	infrastructure of transport on surface
strike of the deposit	mining methods	haulage expenses	energy transfer
inclination angle of the deposit	efficiency of transmission of the object	maintaining expenses	population on the surface of terrain
number of layers or ore bodies and their mutual position	form and quantity of available energy	ventilation and drainage costs	water supply
morphology of the ore bodies	capacity and speed of haulage	transport costs	supplying material for underground mine operation
deposit limit reserves, quality, quantity and distribution of useful component	ventilation system	ore losses as consequence of type of opening	
special geological features, physical, mechanical, hydro--geological, structural	drainage system	time needed for horizon preparation	
tectonics of the terrain and of the deposit	manner of transport and haulage	value of the terrain over the pit	

From the viewpoint of application of transport mechanization the mine may be opened and equipped in such a manner that transport and haulage are executed by one means of transport. In this case, as a rule, are used transport belts and pit trucks. Technological scheme is very simple and effective, namely there is no need for discontinuous transport and reloading of the ore. The construction of the entire system is extremely simple. The restrictions of such an approach are reflected in exploitation depth where it may be effectively applied, the depth is around 300–400 m.

Combined means are applied. One or more kinds of means of transport deliver material to main haulage rooms, where the reloading is done. Haulage is performed by another   kind of means of transport. The scheme has wide application, it is flexible and there is no restriction by depth of occurrence of the deposit. Negative side of this is discontinuous transport, larger number of rooms to be opened and accordingly complex construction of the system.

A great number of factors have influence on the solving of problems during opening. In order to have a more effective insight into the problems they were classified according to their nature and characteristics. According to these criteria four groups have been formed and shown in the table 1.

4. MULTICRITERIUM OPTIMIZATION

Taking into consideration the complexity of the problems that are encountered and solved when mining, it is clear that projects are unavoidably of the multidimensional, and frequently of the controversial nature. Multidimensional nature of mining projects may be made exceptionally simple, and in a certain way it may be distorted if it is described and presented by way of a single criterion.

Theory of management suggests that the consequences of the managing actions always and explicitly aggregate into one criterion function. In case that criteria functions are of similar values relation of indifference is valid. But if there is even a slightest difference then the relation of preference is valid. Such an approach has been explained by the tendency to make comparisons simpler. However, practice refutes this idea.

Abandoning unconditional acceptance of such an approach brought about development of multi-criterial optimization, which in this paper represents the basis for the development of the model for decision support. [1]

4.1. General Terms on Multi-Criterion Optimization (MCO)

Every model of multi-criterion optimization contains the following elements:
− more criteria (target function, criterion function) for decision making;
− more alternatives (solutions) for selection;
− selection process of one final solution.

There are two kinds of MCO problems if viewed from the point of description of the considered reality by means of a mathematical model:
1. multi-target decision making (MTD) or target programming as the subgroup of MCO, and
2. multi-attributive decision making (MAD) or multi-criterion analysis.

It is customary that MTD problems are called well structured, and MAD problems poorly structured.

In order to have a simpler insight into the basic characteristics of MTD and MAD a short comparative analysis is shown in the Table 2.

Table 2

	MTD	MAD
Presence of several criteria defined with	targets, criteria	attributes
Target	explicit	implicit
Attribute	implicit	explicit
Limitations	active	inactive
Alternatives	infinitesimal number of continual	final number discrete
Interaction with DO	distinct	not distinct
Application	designing, finding solutions	selection, evaluation (solutions are known)

Basic terms are similar both for MTD and MAD.

4.2. Basic Principles of Vector Norm

If we assume that in M plain is defined rectangular Descartes coordinate system $(O; \vec{i}, \vec{j})$, then

$$\vec{a} = a_1 \vec{i} + a_2 \vec{j} \quad [2].$$

Now we can calculate the intensity of this vector $\|a\| = \sqrt{a_1^2 + a_2^2}$.
For such defined intensity of the vector is valid:

(i) $\|\vec{a}\| \geq 0 \,\& \, (\|\vec{a}\| = 0 <=> \vec{a} = 0$

(ii) $\|\lambda \vec{a}\| = |\lambda| \|\vec{a}\|, \quad \lambda \in R$

(iii) $\|\vec{a} + \vec{b}\| \leq \|\vec{a}\| + \|\vec{b}\|$ \hfill (1)

Most frequently used vector norms are:

$\|\vec{a}\|_1 = |a_1| + |a_2| + |a_3|$ (l$_1$ norm)

$\|\vec{a}\|_2 = \sqrt{a_1^2 + a_2^2 + a_s^2}$ (Euclides norm)

$\|\vec{a}\|_\infty = max\{|a_1|, |a_2|, |a_3|\}$ (Cebisevljev norm) \hfill (2)

In our case we decided on Euclides norm.

4.3. Entropy

Decision making matrix is as follows:

$$V = \|x_{i,j}\| \quad i = 1,2, \ldots, m \quad j = 1,2, \ldots, n \hfill (3)$$

Normalized decision making matrix is obtained by following equation:

$$R = \|r_{i,j}\|$$

$$r_{i,j} = \frac{x_{ij}}{\sqrt{\sum_{i=1}^{m} x_{ij}^2}} \hfill (4)$$

$J = 1, 2, \ldots, n.$

Criteria importance is a reflection of the decision maker's subjective preference as well as the objective characteristics of the criteria themselves [3]. In order to determine criteria importance, we applied concept of the entropy method. Shannon and Weaver [4] proposed the entropy concept and this concept have been highlighted by Zeleny [3] for deciding the objective weights of criteria. Entropy is a measure of uncertainty in the information formulated using probability theory. To determine weights by the measure, the normalized decision matrix $R = \|r_{i,j}\|$ given by Eq. (4) is considered. The amount of decision information contained in Eq. (4) and associated with each criterion can be measured by the entropy value e_j as shown in Eq. (5):

$$e_j = -k \sum_{i=1}^{m} r_{ij} \cdot \ln r_{ij} \tag{5}$$

where $k = 1/\ln m$ is a constant that guarantees $0 \leq e_j \leq 1$. The degree of divergence (d_j) of the average information contained by each criterion C_j ($j = 1, 2, ..., n$) can be calculated, as per Eq. (6):

$$d_j = 1 - e_j \tag{6}$$

The objective weight for each criterion C_j ($j = 1, 2, ..., n$) is thus given by Eq. (7):

$$w_j = d_j / \sum_{j=1}^{n} d_j \tag{7}$$

Finally the weighted normalized decision matrix is as follows (Eq. (8)):

$$D = \left\| r_{i,j} \cdot w_j \right\| = \left\| \begin{matrix} r_{1,1} \cdot w_1 & \cdots & r_{1,j} \cdot w_j \\ \vdots & \ddots & \vdots \\ r_{i,1} \cdot w_1 & \cdots & r_{i,j} \cdot w_j \end{matrix} \right\| \tag{8}$$

4.4. Algorithm

I step
Defining criteria and decision making matrix.

II step
Calculating criteria weight values (ponders) by method of 'entropy' and adopting ponders on the basis of results of calculations and expert insight.

III step
Calculating vectors norm.

4.5. Example

The opening of a poly-metallic deposit of the vein type has been studied here. Exploitation period is to be 10 years. Annual capacity is 50.000–100.000 t. The deposit has been explored also by mining work and that by vertical shaft and horizontal rooms (drifts) at three levels with difference in height of 40 m. The deposit is at the depth of 100 m (from the surface of the terrain) inclining up to 180 m of depth (prospected part). There is a realistic possibility to incline even deeper. In the concept of deposit opening exploration shaft has the role of the ventilation shaft, and the main haulage room is to be made. Two possibilities have been studied, and they are: mining of a vertical shaft diameter 6 m (Variant 1, further V1) and mining of a ramp of 18 m² crosscut (Variant 2, further V2) [5].

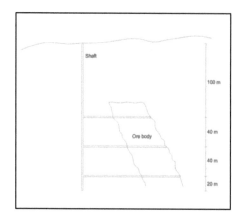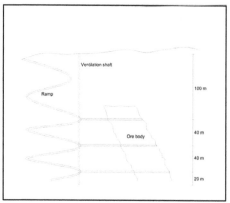

Figure 1. Opening variants: opening with shaft (left), opening with decline (right)

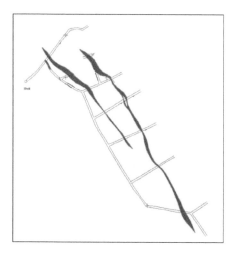

Figure 2. Level 260 of the ore body

I Step – Criteria

k1 expenses for mining rooms for the opening (€/m, €). Only CAPEX are studied. In case of V1 – shaft mining the full amount of expenses has been taken into account due to the assumption that the room will be made within a year. In case of V2 –making the ramp, discounted money flow is taken into consideration. In the first year that is the amount needed for the room length by which the deposit can be reached, later on the lengths are in accordance with average annual advance of the shift (phase opening). Discounting was done by a rate of 10%. More favourable is lower value, i.e. minimizing is done.

Value of the criterion k1

For V1, k1 = 900,000 €

For V2, k1 = 1,384,913 €

Table 3. Discounted money flow of the expenses for ramp constructing

Year	i	ii	iii... x
Price (€)	1,007,174	89,638.51	89,638.51
DCF	1,384,913		

k2 – flexibility means that the room is adaptable to changeable conditions of the deposit. The number is unnamed; it is the result of expert estimation. More favourable is higher value of the criterion (maximizing).

For V1, k2 = 0.4

For V2, k1 = 0.6

k3 – "availability", unnamed number means expert estimation whether the room is suitable for maintaining and use, whether the manner of driving energy supply is adequate, whether available labour has been trained and is able to work on the concrete system, what range of room opening (horizontal rooms) is indispensable for the connecting of main room opening, how effective are ventilation and drainage, etc. More favourable situation is when criteria have higher values (maximizing).

For V1, k2 = 0.4

For V2, k1 = 0.6

Table 4. Decision making matrix

	k1 (min)	k2 (max)	k3 (max)
V1	900,000	0.4	0.4
V2	1,384,913	0.6	0.6

Opening variants parameters:

Table 5. Variant V1

Opening depth(m)	200
Depth (length) of the room (m)	200
Unit price (€/m)	4500
Total price (€)	900,000.00

Table 6. Variant V2

Opening depth (m)	180
Length of the room (m)	1,511
Inclination of the room 12%	12
Unit price (€/m)	1,200
Total price (€)	1,813,921

II Step – Weight Coefficient (Ponders)

Normalized decision matrix (4)

$$R = \begin{vmatrix} 0,5449 & 0,5447 & 0,5447 \\ 0,8384 & 0,8321 & 0,8321 \end{vmatrix}$$

The entropy:

$$E_j = |0{,}3316 \quad 0{,}3324 \quad 0{,}3324|$$

$$d_j = |0{,}6684 \quad 0{,}6676 \quad 0{,}6676|$$

$$w_j = |0{,}3335 \quad 0{,}3332 \quad 0{,}3332|$$

$$w_1 = 33.35\%$$

$$w_2 = 33.32\%$$

$$w_3 = 33.32\%$$

Such result shows that all criteria (as they are defined) have almost similar ponder. However the estimation is that this should not be the case, i.e. in this way the expert preferences or those of decision makers would not be taken into account. There are several ways to correct ponders. Here we mention two. The first one is the so called hybrid ponder that is arrived at by combination of expert estimation and calculated ponder. The value is calculated on the basis of the formula:

$$w = \frac{w_0 w_j}{\Sigma w_0 w_j} \tag{9}$$

The second one is by expert estimation. Here we have decided on the second. So, our p onders have the value:

Table 8. Ponder values

w_1	0.50
w_2	0.25
w_3	0.25

In this way we emphasize the criterion that we think should have the greatest influence on the final decision.

III Step – Calculating Norm Vectors

Norm vectors are calculated [Petrasov 1978] according to the algorithm:

$$\Delta k_{ij} = \frac{k_{ij} - k_{ijo}}{k_{ijo}} \tag{10}$$

$$R_i = \sqrt{x_1^2 w_1 + x_2^2 w_2 + \cdots + x_n^2 w_n} \tag{11}$$

The smallest value of the norm vector represents a more favourable variant.

R1	0.204124
R2	0.367012

In this case it is obvious that it is variant 1 (V1), i.e. opening by shaft.

CONCLUSION

The presented model showing the estimation of the opening system represents a tool for decision support. Subjective estimations given by experts are quantified via criteria k2 and k3, as well as by adopting weight coefficients – ponders.

The model is, as can be seen "open", i.e. adjustments to different expert estimations are possible. The number of criteria is not limited. According to our opinion the model for decision support is simple and comprehensible both for the one who is estimating as well as for the decision maker, thus in a certain way the "black box" effect that is so much present today, is being avoided.

ACKNOWLEDGEMENT

The paper is part of research conducted on scientific projects TR 33003 that are funded by Ministry of Science and Technological Development, Republic of Serbia.

REFERENCES

[1] Petrović R. 1986: Upravljanje sistemima. Naučna knjiga, Beograd.
[2] Scitovski R., Brajković D. 2013: Geometrija ravnine i prostora.
    http://www.mathos.unios.hr/geometrija/Materijali/Geo_1.pdf
[3] Zeleny M. 1982: Multiple Criteria Decision Making. McGraw Hill, New York.
[4] Shannon C.E., Weaver V. 1947: The Mathematical Theory and Communication. The University of Illinois Press, Urbana, Illinois.
[5] Petrasov A.A. 1978: Modelirovanie i optimizacija processov na rudnikah. Nedra, Moskva, IB No. 1432.

International Mining Forum 2015, Kicki et. al. (eds) © 2015 Taylor & Francis Group, London, UK. ISBN 978-1-138-02820-3

Two-Dimensional Raise-Bore Hole Stability and Support Assessment for the Victoria Project in Ontario, Canada

M. Jakubowski, M. Moskwa
KGHM Polska Miedź S.A.

ABSTRACT: The paper deals with the geotechnical assessment of raise-bore hole stability using empirical and numerical stability analyses for the Victoria Project in Ontario, Canada. The scope of work for this paper presumed the raise bore is to be excavated to a 4,0 m diameter and to an approximate depth of 1860 m. The geotechnical and structural data, along with the interpreted major lithological units from the exploration shaft pilot hole were used to evaluate the rock mass classification parameters for the raise-bore hole.

KEYWORDS: Stability, Raise-Bore Hole, Main Access Openings, Support, Reinforcement

1. INTRODUCTION

The Victoria Project is a proposed underground copper-nickel mine located in the Sudbury mining camp, an environment historically mined for over 110 years for its sulphide nickel, copper, cobalt and precious metal ore bodies. The Victoria Property is located approximately 35 km by road from the downtown core of the City of Greater Sudbury in northeastern Ontario, Canada, and approximately 400 km north of Toronto, Ontario.

The key deliverables for the Project are construction of the surface infrastructure, sinking of the exploration shaft, underground definition drilling, extraction of a bulk sample, and the underground infrastructure to support production. For the exploration purposes, it is necessary to sink a vertical shaft down to the level 1830 below the surface. Although only one shaft would be required to access and complete the diamond drilling program, two main openings would be required to mine the zone. In order to support a full mining production from the orebody either the second vertical shaft or the raisebore holes are crucial to be a second egress and to be used as a main ventilation route.

2. GEOLOGICAL SETTING

The Victoria Property is part of the Sudbury Structure for the greater than 15,000 km^2 area represented by the sedimentary and brecciated rocks of the Whitewater Group (Onaping, Chelmsford and Onwatin Formations), the elliptically shaped Sudbury Igneous Complex (SIC) and the surrounding brecciated country rocks of the Superior and Southern Provinces. This unique geological environment is situated along a zone of Early Proterozoic faulting and dislocation known as the Murray Fault System.

The footwall rocks on the north and east margins of the SIC are the Archean Levack Gneiss Complex and granitoids. The footwall rocks to the south are Paleoproterozoic Huronian Supergroup metavolcanic and metasedimentary rocks. A summary of major lithological units as interpreted is illustrated on Figure 1.

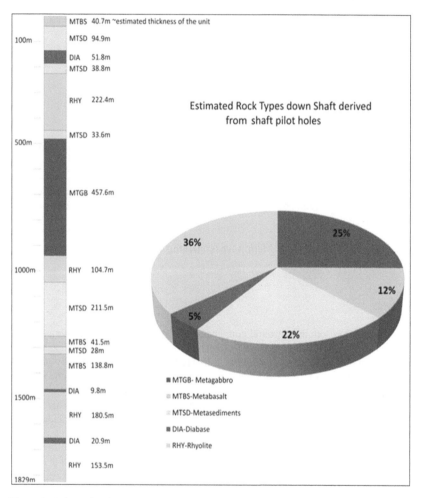

Figure 1. Estimated rock types by depth along the Victoria exploration shaft

3. EMPIRICAL RAISEBORES STABILITY ASSESMENT

Methods for evaluating shaft and raise-bore hole stability empirically are limited in the literature. The McCracken and Stacey empirical shaft stability method (1989) is a method primarily used for evaluating wall stability in raise-bore shafts. This is used as a starting point for the evaluation of anticipated wall stability for the Victoria ventilation raise-bore hole. The basic concept of M&S is to classify the rock mass using Q [Barton 1974] for each domain of interest, and apply a number of adjustment factors to obtain a raise-bore quality index Q_R. First an adjustment to Q for the side walls ($Q_{sidewall}$) is applied as follows:

$$Q_{sidewall} = \begin{cases} 2.5 \ where \ Q > 1 \\ Q \ where \ Q \leq 1 \end{cases}.$$

Next, an orientation adjustment for the effect of major discontinuity sets on the walls is made ($A_{orientation}$) according to the adjustment factors listed in Table 1.

Table 1. Discontinuity orientation adjustment factors

Number of steeply dipping (60–90°) major joint sets	$A_{orientation}$
1	0.85
2	0.75
3	0.60

Finally, a weathering adjustment is made based on the likelihood of weathering of the rockmass and its degree in consideration of its effect on stability of the shaft over the long term is made. The adjustments for weathering ($A_{weathering}$) are 0.9, 0.75 and 0.5 for slight, moderate and severe degrees of weathering respectively of the intact rock of the sidewall over the long term. The adjustments are applied cumulatively to obtain Q_R:

$$Q_R = Q_{sidewall} \cdot A_{orientation} \cdot A_{weathering}.$$

The raise-bore stability ratio (RSR) is a term used to describe the purpose of the excavated shaft, criticality and its intended life for use. M&S suggests an RSR of 1.3 is appropriate for a ventilation shaft. A RSR = 1 has been assumed for this permanent and critical infrastructure for the mine. A raise-bore stability ratio (RSR) is then evaluated based on the dependence of the shaft and its service life. The Q_R is then plotted on the raise diameter vs. Q_R chart along with the applicable RSR to determine the maximum unsupported diameter of the shaft (Fig. 3).

Table 2. Q classification and M&S raise bore quality index Q_R data summary [Hume & Kalenchuk 2014]

Rock Type	Q'$_{joints}$	Jw	SRF	Q	Q$_{sidewall}$	A$_{orientation}$	A$_{weathering}$	Q$_R$
Depths from surface to 500 m								
MTSD	3.8-12.5	0.66-1	2.5	0.8-5.0	0.8-12.5	1	0.75-1	0.7-12.5
MTBS	5.6-12.5	0.5-1	2.5	1.1-5.0	2.8-12.5	1	0.5-1	1.4-12.5
RHY	7.5-12.5	1	2.5	3.0-5.0	7.5-12.5	0.75	0.9-1	5.1-9.4
DIA QDIA MDIA	3.8-12.5	1	2.5	1.5-5.0	3.8-12.5	0.85	0.9-1	2.9-10.6
Depths from 500 m to 1200 m below surface								
MTGB MXGB	11.3-12.5	1	6-9	1.3-2.1	3.1-5.2	0.75	1	2.4-3.9
MTSD	3.8-12.5	1	6-11	0.2-2.1	0.2-5.2	1	1	0.2-5.2
RHY	7.5-12.5	1	5	1.5-2.5	3.8-6.3	0.75	1	2.5-4.7
Depths greater than 1200 m below surface								
MTBS	5.6-12.5	1	14	0.2-0.9	0.2-0.9	1	1	0.2-0.9
RHY	7.5-12.5	1	13	0.6-1.0	0.6-1.0	0.75	1	0.4-0.7
DIA QDIA MDIA	3.8-12.5	1	15	0.3-0.8	0.3-0.8	0.85	1	0.2-0.7

The evaluation of Q is made by determining Jw and SRF, and multiply the Q' value listed in Table 2 by Jw/SRF. A hydrogeological testing program was performed in 2012 and results of the field investigations performed at depths of 31m or greater below surface demonstrated very low hydraulic conductivity and concluded There are no potential issues with significant water inflow in the proximity of pilot hole. At depths less than 31 m, MTBS (0–40 m) has been assigned assuming the potential for significant rainfall events, spring melts and higher permeability due to the higher weathering of rock near surface (Jw = 0.5–1). MTSD has been assigned assuming the potential for medium inflow (Jw = 0.66–1). For depths that are greater than 100m, it is assumed that the shaft will be dry with minor inflow (Jw = 1).

As can be seen in Figure 2, Victoria raise-bore hole diameter of 4.0 m both plot beyond the entire range of case histories reported from Coombes et al. (2011) at all depths. At this diameter, Q_R for all lithological units plot outside of, and are well above the "unstable" limit of the RSR = 1 line. This highlights the need for raise-bore hole to be fully supported. It must be well understood that the M&S method is intended for raise-bore shafts and holes, which are assumed to have minimal disturbance to the rock mass immediately adjacent to the shaft walls. If the given shaft or large diameter hole is to be sunk conventionally using drilling and blasting, it is expected that this will disturb and weaken the rock mass.

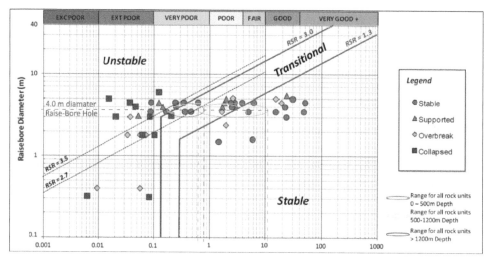

Figure 2. Victoria 4.0 m raise-bore hole, and range of Q_R for all major lithological units plotted on the raise-bore diameter vs. Q_R chart [Jakubowski, after Coombes et al., 2011]

4. ANALYTICAL RAISEBORE STABILITY AND SUPPORT ASSESMENT

Two-Dimensional finite element analysis using Phase2 v. 8.0 (software by RocScience™) has been conducted at various depths to assess potential stress related stability concerns within the raise-bore hole. The major lithological units have been assessed at various depths in accordance with the interpreted units and their depths for the exploration shaft as given in Figure 1. The vertical raise-bore hole advance is simulated in 2D as a cross-section at the varying depths as plane-strain using the material softening approach. The method simulates the advance of the shaft by replacing the excavated material with an internal pressure equal in magnitude but opposite in direction to the hydrostatic stress field on the excavation boundary. This pressure is reduced incrementally down to zero, representing the fully excavated state.

To design a support system, it is necessary to determine the amount of the excavation wall deformation prior to support installation. As a hole is excavated, there is a certain amount of deformation before the support can be installed. Determining this deformation can be done using either observed field values, or numerically from 3D finite-element models or axisymmetric finite-element models, or by using empirical relationships such as those proposed by Panet or Vlachopoulos and Diederichs. Using the core replacement technique, it is required to determine the modulus reduction sequence that yields the amount of tunnel wall deformation at the point of and prior to support installation. At the end, it is possible to build a model that relaxes the boundary to the calculated amount of deformation. The support should be added to determine if the excavation is stable, if the excavation wall deformation meets the specified requirements, and if the excavation lining meets certain factor of safety requirements.

The joint set incorporated into the numerical model is based on stereonet plots and borehole televiewer data from the shaft pilot hole FNX1204 (Fig. 3). The in-situ stresses are estimated at depth using the equations listed in Table 3 from Trifu & Suorineni (2009). Z is depth in metres. The major principal stress $\sigma1$ and intermediate $\sigma2$ are assumed to be horizontal, with $\sigma1$ oriented parallel to the ore body strike.

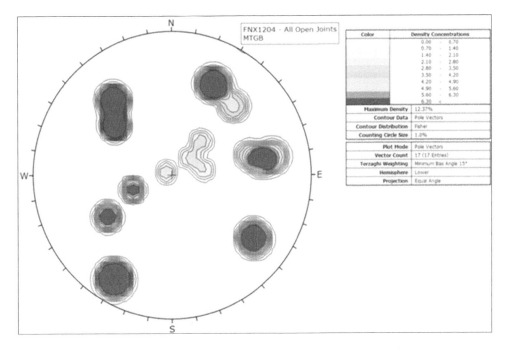

Figure 3. Stereonet plot of all open discontinuities in Metagabbro (Golder Associates Ltd., 2012)

29

Table 3. Principal stress magnitude calculation [Trifu & Suorineni 2009]

Principal Stress	Magnitude (MPa)	MTGB layer principal stresses (MPa)
σ_1	10.9 + 0.0407 Z	47.53
σ_3	8.7 + 0.0326 Z	38.04
σ_z	0.029 Z	26.10

Table 4. Basic MTGB Material Properties incorporated to the FEM model (Jakubowski)

Parameter	Value	Unit
Young's modulus	20901.6	MPa
Poisson's ratio	0.3	-
Compressive strength	48.2	MPa
mb parameter	19.578	-
s parameter	0.367879	-
a parameter	0.500174	-

The aim of the first stage of the numerical modeling was to estimate the deformation of the wall which occurs before any liner is installed. The material softening technique was applied. Next, the screening was utilized as an initial lining to find out its influence on the strength factor and walls displacement. Because of limited influence of rock bolts, the 10 cm thick concrete liner was applied (Tab. 4). Total displacement of the raise-bore walls sunk in MTGB layer is shown on Figure 4, while Figure 5 illustrates the factor of safery envelopes versus the shear forces space and versus the moment space.

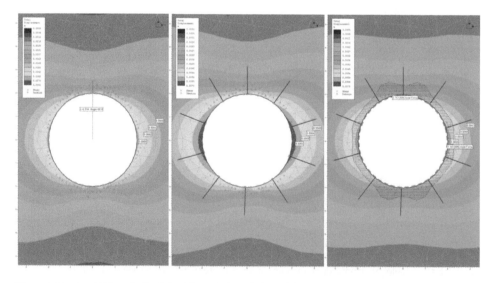

Figure 4. Results of 2D analysis of the 4.0 m raisebore hole sunk in MTGB layer (Jakubowski)

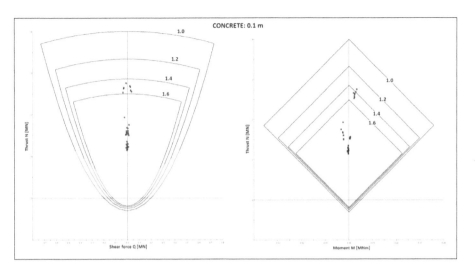

Figure 5. Support capacity diagrams (Jakubowski)

CONCLUSIONS

It is very safe to say that the raise-bore technology has improved significantly over the last decade, and will improve further by the time raises at Victoria are required. However, with increasing mining depth and longer raise-bore holes increases the potential for squeezing while reaming. What is more, increasing the length of a raise-bore also increase the likelihood of intersecting adverse ground conditions. Given the initial unsupported nature of a raise-bore technology, this poses a potential problem that needs to be thoroughly understood.

In the numerical modeling presented in this paper the raise-bore hole that intersected the MTGB layer indicated high walls displacement (6,8 mm) caused by horizontal stresses. Screening as an initial support did not change the geomechanical stability (strength factor) but slightly limited the displacement. After rock bolts installation the total displacement was estimated to 6,5 mm. Once the 0,1 m thick concrete lining was applied, the displacement dropped rapidly to 5,9 mm. The combined bolt-concrete liner indicates the factor of safety higher than 1.2 and no element yielded. However, it is suggested to further examine thicker concrete support or reinforced concrete support as well as the liner installation sequence. This should limit the wall displacement, improve the strength factor and increase the factor of safery.

REFERENCES

Barton N., Lien R. & Lunde J. 1974: Engineering Classification of Rock Masses for the Design of Tunnel Support. Rock Mech., 6, 183–236.

Coombes B., Lee M. and Peck W. 2011: Improving Raisebore Stability Assessments and Risk. In Proceedings of Mines without Borders, CIM Conference and Exhibition, Montréal, 22–25 May 2011, Canadian Institute of Mining, Metallurgy and Petroleum (CIM), Montréal, Que.

Golder Associates Ltd. 2012: Factual Report of Geotechnical Study Shaft Pilot Hole (FNX1204). Report Number: 10-1193-0018, March 30, 2012.

Hume C. 2011: Numerical Validation and Refinement of Empirical Rock Mass Modulus Estimation. M.Sc. Thesis, Queen's University, Kingston.

Hume C. and Kalenchuk K. 2014: MDEng Technical Memo #0308-M1410-02: Summary of 8.3 m Shaft Stability Assessment – Victoria Project – KGHM International Ltd.

McCracken A. and Stacey T.R. 1989: Geotechnical Risk Assessment for Large Diameter Raisebored Shafts. IMM Proceedings of Shaft Engineering Conference, pp. 309–316.

Trifu C. and Suorineni F. 2009: Use of Microseismic Monitoring for Rockburst Management at Vale Inco Mines. In: Controlling Seismic Hazard and Sustainable Development of Deep Mines, C. Tang (ed.), 1105––1114.

International Mining Forum 2015, Kicki et. al. (eds) © 2015 Taylor & Francis Group, London, UK. ISBN 978-1-138-02820-3

Conditions of Deformation Prediction in the Shaft

A. Kwinta
Faculty of Environmental Engineering and Land
Surveying University of Agriculture in Krakow

ABSTRACT: Mine shafts shall function safely according to their destined use over the entire time of exploitation. In order to secure the shaft function protection pillars are established around it within which pillars mining is not allowed or only on special conditions. Forecasting a continuous shaft deformation due to mining under Polish conditions is carried out by means of the modified Knothe theory. Due to various modifications of theory and its parameters are changes the quality of the prediction. In the paper was presented different functions of influence radius in rock mass.

KEYWORDS: Deformation prediction, deformation model, shaft deformation

1. INTRODUCTION

In an era of reducing mining costs, companies are more and more often looking at the reserves located near the main excavations i.e. near the shafts. In Polish mining, a protective pillar is set up in the area of the shaft, whose job is to ensure shaft safety. Undoubtedly, such a measure is essential for the safe operation of the mine. On the other hand, the amount of deposits trapped in the pillar is significant, and also taking into account the short transport route of excavated material, mining in the area of the shaft is very attractive economically. The purpose of the shaft pillar is to move the influence of mining away from the pipe shaft, but sometimes mining is carried out inside the pillar, and in such cases theoretical calculations should be used to determine the deformations that may occur [Majcherczyk et al. 2003]. Consequently, a decision must be made to allow the implementation of the given mining operations or to reduce them.

Forecasting of deformation is possible if a model of the occurrence of deformation adequate to reality is adopted and the proper set of its parameters is available [Niedojadło 2008]. Beginning in the nineteenth century, people started to record the formation of subsidence troughs on the surface of the earth using geodetic methods. Based on observations, construction of models was attempted that would allow forecasting deformation. Currently a theory belonging to the group of integral geometric theories is being widely used for the calculation of deformation. This theory was originally developed for the surface, and only later was developed for the rock mass and expended by adding a time function [Knothe 1984]. And so, the Knothe theory for the calculation of deformation within the rock mass is a secondary method [Collective work 1980]. Distribution of deformation indices within the rock mass is determined by the selection of the appropriate scope and timing of mining, so as to minimise damage to the pipe shaft (e.g. [Dżegniuk 1967], [Piwowarski et al. 1995], [Collective work 1980]), but it also depends on the adopted theoretical model and its parameters.

In the case of calculations for the rock mass in the mining area, geomechanical methods with the designated rock mass parameters are used [Majcherczyk et al. 2013], in practice, much simpler geometric methods have passed the test for calculations in shafts [Piwowarski ct al. 1995]. However, even in the case of integral geometric methods for forecasting the deformation, selection of the appropriate influence function and a set of parameters adequate for the mining and geological situation is important. For the surface of the area, the form of the influence function and the set of parameters is limited, but for the rock mass (the shaft), it is necessary to take into account the propagation of deformation in the influence function of the theoretical model.

Below, a number of comments related to the forecasting of deformation along the pipe shaft are presented. Over the decades I have been the author or co-author of scientific studies on the forecasting of deformation along the pipe shaft in both coal mining, salt mining, and copper mining. For the calculation of the deformation, I use programs, which I have authored or co-authored.

2. DEFORMATION FORECASTING PROCEDURE

Calculation of the deformation caused by mining consists of a series of operations, from the selection of forecasting theory, through preparing the data, performing the calculations, up to the final report and the analysis of results. Currently, it is possible to make calculations for a large number of points in multiple time horizons, taking into account hundreds and sometimes thousands of mining areas [Hejmanowski, Kwinta 2009]. Calculations of the deformation for large "projects" can be done quickly, and the only problem is the correct interpretation of the results. The introduction of deformation forecasting systems allowed for thc automation of the calculation process, while facilitating the management of data and calculation results. Many different programs for forecasting deformation have been developed, but it seems that the best tool for this type of operation are databases, which make it possible to easily acquire and collect the data necessary for calculation, and to process deformation values obtained in the calculations.

Regardless of the theoretical calculation methods used to forecast deformation, the following stages of calculation must be completed:

1. Selection of the forecasting model determines all subsequent calculation steps. Of course, the key issue is to select a model that will be adequate to the realities in the given circumstances. If the model has already been used in the mining plant and has proven itself, it is an important prerequisite for its use. In the Polish mining industry, the Knothe theory, with various modifications, is commonly used to forecast deformation.
2. Preparing data for the calculations is primarily associated with identifying the scope of the calculations in terms of space and time. Distribution of calculation points and the choice of time horizons are directly related to the shaft. Points should be spaced regularly along the pipe shaft and additionally in characteristic points (e.g. shaft bottom, locations of damage to the pipe shaft). It is also necessary to include the technical infrastructure of the shaft located on the surface in the calculations. Time horizons are associated with shaft inspections or repairs, dates of measurements, as well as the schedule of mining in the shaft area. The distribution of points and the choice of time horizons affect the reserves in the mining fields, which should be included in the calculation. Data related to the mining operations carried out is obtained from mining maps (geometry, depth, thickness, progress of mining), and for the proposed mining operations, data are taken from the Operations Plans, Deposit Management Plans or different variants developed in the output preparation department.
3. Configuring the calculations for the adopted model involves the selection of model parameters relevant to the given mining conditions. In the classic Knothe theory, there are two parameters for asymptotic states. Modifying the theory by adding the ability to perform calculations within the rock mass, in time, or taking into account a change of the properties of the medium (slope, anisotropy) causes that the calculations may have several parameters whose values must be determined [Hejmanowski 2001].

4. Fundamental theoretical calculations including the implementation of the calculation algorithm designed to determine the basic deformation indices. Currently, all calculations are carried out with the use of computers, and now there are even applications for mobile devices like tablets or smartphones. Calculations can be performed in applications or systems that are connected to databases, where deformation indices will be processed [Hejmanowski, Kwinta 2009]. Integral geometric models are implemented by calculation algorithms using numerical integration.
5. Besides the classic deformation indices, there is sometimes a need for determining their processed values [Kwinta 2012] (e.g. value increase, speed, main deformation), or take account of other factors that cause constant deformation (e.g. rock drainage). In this case, it is necessary to perform additional calculations. Ideally such calculations should be performed based on a database, in which case the calculations are performed quickly while minimising the possibility of errors.
6. A deformation calculation report should include a graphical summary (chart, distribution map), figures (extreme values of indices, tables), a description covering the characteristics of the data included in the calculation, a brief description of the computational model with the values of its parameters, the analysis of the obtained values of deformation indices and any recommendations for the implementation of the proposed mining operations, conducting surveys, inspections of the condition of the shaft.

Selection of the computational model with its parameters significantly determines the quality of the results. Deformation forecasts can be verified by comparing the calculation results with the indicators calculated on the basis of surveying.

3. COMPUTATIONAL MODEL

As previously mentioned, the current basic computational model for deformation indices is the Knothe theory with later extensions and modifications [Hejmanowski 2001]. Virtually every specialist in forecasting deformation, especially in the rack mass and in time, has developed their own calculation formulas. The following are the basic formulas and some modifications that are relevant to the calculation procedure along the pipe shaft.

The integral geometric theories for forecasting deformation assume that the elementary mining in the volume of deposit dV having a horizontal surface dP is accompanied by elementary subsidence at point A, remaining in the range of impact of this mining operation. This subsidence can be described as follows:

$$dw_A = f(x, y)dP \tag{1}$$

where: dw_A – elementary subsidence in point A, $f(x, y)$ – influence function, dP – surface of elementary mining.

Using the principle of superposition, it is assumed that the vertical displacement at point A is the sum of the elementary subsidence coming from all the elementary volumes of the mining field:

$$w_A = \iint_P f(x, y)dP \tag{2}$$

By adopting various forms of influence functions $f(x, y)$ we obtain different theories for predicting displacements and deformations. In his theory, Knothe adopted the influence function based on

35

a Gaussian function [Knothe 1984]. The formula for the influence function in the case of flat displacement in the Knothe theory takes the form (3):

$$f(x) = w_{max} \frac{1}{r} exp(-\pi \frac{x^2}{r^2})$$
(3)

where: $w_{max} = a\,g$ – the maximum final subsidence [m], a – exploitation factor, dependent on the way of filling the abandoned workings, g – thickness of exploitation, x – distance from mining boundary, r – radius of influence dispersion.

Size r is the radius of influence dispersion, which is also called the radius of the range of main influences. This volume is related to the characteristics of the medium in the form of the angle of the range of main influences, via the formula:

$$tg\beta = \frac{H}{r}$$
(4)

where H is the depth of the field [m].

Using the classic Knothe theory to determine the indicators, calculations can be carried out only for the land surface. According to formula (4), radius of the range in the classic theory is determined only for the land surface.

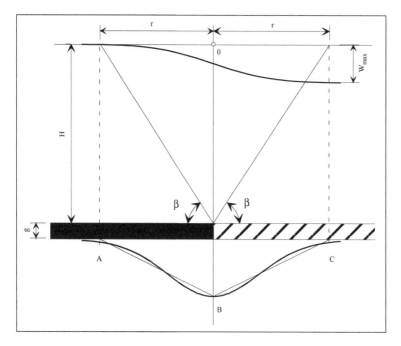

Figure 1. The parameter of influence dispersion r and the angle of the reach main influences β [Knothe 1984]

In the literature, one can find a whole range of functions describing the passage of deformation through the rock mass based on the variability within the radius of influence range in the rock mass. This issue has been dealt with, among others, by Knothe (1953), Budryk (1953), Drzęźla (1979), Kowalski (1985), Jędrzejec (1986). Due to the lack of a corresponding set of geodetic measurements inside the rock mass, work on the construction of the function $r(z)$ is mainly based on theoretical considerations [Budryk 1953] and model studies[Litwiniszyn 1962]. Some attempts to verify the results obtained were conducted in random observational material [Kowalski 1984]. Analysis of the different forms of the range radius function in the rock mass can be found in the literature [Kwinta 2009], [Niedojadło 2008].

Several characteristic solutions are presented below:
– Knothe equation (1953):

$$r(z) = r(H)\frac{z}{H} \tag{5}$$

– Budryk equation (1953):

$$\frac{r(z)}{r(H)} = \left(\frac{z}{H}\right)^n \tag{6}$$

where $n = \sqrt{2\pi}tg\beta$.

– Drzęźla equation (1979):

$$r(z) = r(H)\left(\frac{z + z_0}{H + z_0}\right)^n \tag{7}$$

where: $0.405 > n > 0.735$ – most often entertains one another $n = 0.6$, z_0 – parameter of modelling the radius of influence dispersion on heights of the exploitation.

Figure 2 summarizes the form of the functions of variation in the range of main influences depending on the chosen solution.

Forecasting deformation indices in the horizontal plane became possible using integral geometric theories thanks to Awierszyn's (1947) observation. This author, analyzing the rich empirical evidence and making theoretical deliberations, determined the relationship between horizontal deformations and curves in the form of the following differential equation (8):

$$\frac{dU}{dx} - B\frac{d^2w}{dx^2} = 0 \tag{8}$$

The proportionality factor B appearing in this equation has dimension [m], and thus Awierszyn (1947) sought an explanation for this factor in the rock mass. The author suggested an interpretation that B is a distance between the surface and the neutral axis of the yieldable layer.

Budryk's (1953) deliberations were important for the prediction of horizontal displacements and deformations in Poland. By introducing a modified Awierszyn formula, Budryk expended the Knothe theory by the possibility of determining the deformation indices in the horizontal plane. Budryk obtained the following form of the formula for coefficient B:

$$B = \frac{r}{\sqrt{2\pi}} \approx 0.4r \tag{9}$$

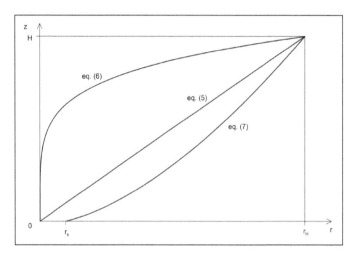

Figure 2. Shape of model functions of the changeability of the radius
of influence dispersion in rock mass according to equations (5)–(7)

It should be noted that formula (9) was derived for the surface area. To determine the values of the deformation indices within the rock mass, Budryk proposed formula (6) for the form of the radius of influence in the rock mass. Currently, this formula usually takes the form:

$$B(z) = b \cdot r(z) \tag{10}$$

where b is an appointed parameter.

The models presented so far that allow for performing theoretical calculations for stationary deformation. In reality, mining operations develop over time; therefore the calculations must take into account the time factor. Among the many existing solutions, the most common ones will be presented:

– Knothe formula (1984):

$$\frac{dw(t)}{dt} = c \cdot \left[w^k(t) - w(t) \right] \tag{11}$$

where: c – time coefficient, $w^k(t)$ – final subsidence from stopped mining in the t moment, $w(t)$ – subsidence in the t moment.

– Sroka-Schober formula (1987):

$$\Delta w(t) = a \cdot \Delta V \cdot f(t) \tag{12}$$

where: $\Delta w(t)$ – elementary volume of the subsidence trough in the time t, a – exploitation coefficient, ΔV – elementary volume of the exploitation, $f(t)$ – time function.

$$f(t)=1+\frac{\xi}{c-\xi}\,exp(-ct)-\frac{c}{c-\xi}\,exp(-\xi\,t) \tag{13}$$

where: c – time coefficient for the process of proceeding from exploitation to computational horizons, ξ – time coefficient for process of the convergence of exploitation field.

The theoretical solution presented above does not exhaust the issue of forecasting displacements and deformations caused by underground mining operations. But they are an approximation of the issue.

Summing up the above summary information, when calculating the deformation indices along the pipe shaft, various forms of the influence function with various additions can be taken into account. Below is a set of parameters to be considered in the course of the calculation of deformation indices along the pipe shaft. The values of parameters are adopted for calculations on the basis of the previously implemented mining operations, according to information gathered from the mine and taking into account the average values of the parameters for similar mining operations of this type of deposit:

- $tg\ \beta\square\square\square$ – tangent of the angle of the range of main influence in the Knothe theory (if rock anisotropy is known, it is possible to introduce various values of this parameter into the calculations, in different directions),
- a – parameter dependent on the mining system (method of roof control),
- $b\ \square$ – the coefficient of proportionality between the vertical and horizontal deformation indices,
- $\mu\square\square\square$ – the parameter describing the deviation of the trough due to the slope of the rock layers,
- $\xi\square\square\square$ – time function parameter describing the process of convergence of the selected element of the mining field,
- $c\ -\square$ time function parameter describing the transition of deformation from the mining field to the computing horizon,
- $n\ -\square$ the parameter determining the variability within the radius of the range of main influences in the rock mass, in the area of the analysed bed,
- $r_s\ -\square$ radius of the range of main influences on the level of the roof of the mined bed.

A set of 8 parameters for the computational model, which are used in the calculation of deformation indices, have been presented above. Proper selection of the computational model and its parameters is the basis of expert knowledge. Poorly assumed parameters may result in erroneous values of deformation and the consequent threat to the functioning of the shaft and the mine.

4. THEORETICAL EXAMPLE

To illustrate how the adoption of the variability function within the radius of the range of influences in the rock mass affects the formation of deformation indices along the pipe shaft, a theoretical example was prepared. The calculations use the "MODEZ 4" deformation forecasting system. The illustration shows the calculation diagram. Single longwall mining is planned, which will affect the shaft (Fig. 3).

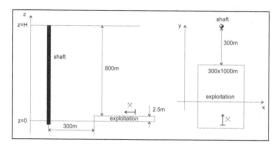

Figure 3. Scheme of the exploitation and
the mining shaft for theoretical example

As seen in Figure 4, depending on the assumed function for the radius of the range of influences within the rock mass, the distribution of deformations along the pipe shaft significantly changes for the assumed example. The use of the function proposed by Drzęźla (7) causes a greater range of influence of mining on the pipe shaft than in the case of other formulas. The formula proposed by Budryk (5) results in the forecast of large vertical deformation values for the points located in the shaft top area.

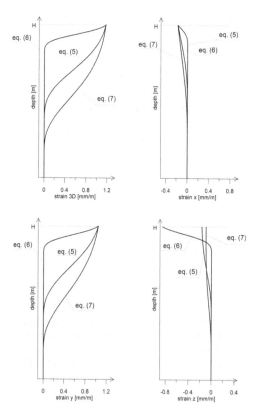

Figure 4. Schedule of deformations for the function of the radius
of influence dispersion in rock mass according to equations (5)–(7)

Hence, it should be noted that, taking into account the safety of the operation of the mine and the shaft, the best solution is to use the function proposed by Drzęźla (7) in the calculations, in case of mining the protective pillar. This results in the adoption of a certain margin of error in the calculation of deformation indices due to the value of deformations forecast along the pipe shaft.

5. PRACTICAL EXAMPLE

Results of calculations of indicators of the deformation appointed in line Drzęźla (7) function of the radius of influence dispersion will be illustrated on the example of analysis of the done and designed coalmine. Of all of mining exploitation were isolated parcels, which ones can an influence on two shafts (Fig. 5). With the whole they chose 141 parcels of the effected done exploitation and 91 parcels of the planned exploitation. The depth fluctuated between 200 and 1020 meters, thicknesses of the mine are from 1.0 into 4.3 m.

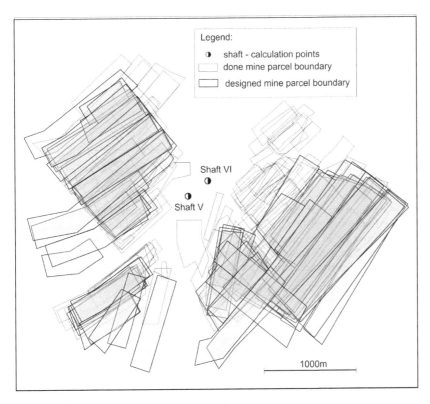

Figure 5. Scheme of the exploitation and mine shafts

Theoretical calculations were carried out with the aid of the deformation forecasting system "MODEZ 4". Parameters for the model of forecasting were assumed according to the Knothe model application for mines. Calculation points were assumed along the profile of V and VI shafts, spaced at 10 m, thus resulting in 180 calculation points.

For each of objects (V and VI mineshaft), out of all calculated deformation, for every calculation point, deformations were chosen in the asymptotic state of the deformation. Results of calculations were presented in Figure 6.

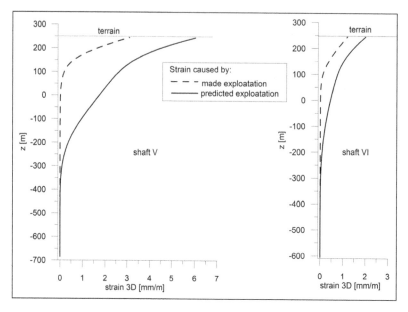

Figure 6. Results of calculations of the maximum 3D deformations along mine shafts

Obtained results are showing that the value of the maximum 3D deformations will increase as a result of the planned exploitation. In case of the V mine shaft values of deformations slightly crossing 6 mm/m are being get in the area of terrain, however for the VI mineshaft the maximum deformations will exceed 2 mm/m. These results of calculation are special interesting for objects located on the terrain near of shaft (e.g. mining headframe).

SUMMARY

Maintaining the functionality of the main mine workings is one of the fundamental issues related to the safety of mining activities. Particular attention in the design of mining activities should be paid to the possible impact of mining on mine shafts. Protective pillars are set up in the areas of shafts, within which the possibility for mining is very limited. However, it is possible to design mining operation to reduce its impact on the shaft. In this case, it is necessary each time to carry out a detailed analysis of the current influences of mining operations carried out so far, analysis of the technical condition of the shaft lining and infrastructure.

Calculations of deformation indices should be carried out with due diligence, taking into account a large amount of data and parameters included in the calculation. An important element is the analysis of the main deformations in the spatial state of deformation, which allows for a full determination of threat to the shaft from the mining operations conducted. The widespread use of computer systems to forecast deformation allows for determining the main spatial deformation.

Increasingly, the attempts are made of mining the deposit in the immediate vicinity of the shafts. This is economically viable and technically feasible, but each case of such mining operations must be carefully examined and, if there are any doubts in regard to the mining operations, they must be abandoned or limited.

REFERENCES

Awierszyn S.G. 1947: Sdwiżenije gornych porod pri podziemnych razrabotkach. Ugletiechzdat, Moskwa.

Budryk W. 1953: Wyznaczanie wielkości poziomych odkształceń terenu. Archiwum Górnictwa i Hutnictwa, t. 1, z. 1, Warszawa.

Collective work 1980: Ochrona powierzchni przed szkodami górniczymi. Wyd. „Śląsk", Katowice.

Drzęźla B. 1979: Zmienność zasięgu wpływów eksploatacji w górotworze. Przegląd Górniczy, nr 10, Katowice.

Dżegniuk B. 1967: Odkształcenia pionowe rury szybowej przy eksploatacji filarów szybowych. Prace Komisji Nauk Technicznych, Górnictwo, z. 5, Kraków.

Hejmanowski R., Kwinta A. 2009: System prognozowania deformacji MODEZ. Materiały konferencji "X Dni Miernictwa Górniczego i Ochrony Terenów Górniczych", Kraków.

Hejmanowski R. 2001 (ed.): Prognozowanie deformacji górotworu i powierzchni terenu na bazie uogólnionej teorii Knothego dla złóż surowców stałych, ciekłych i gazowych. Wyd. Inst. Gosp. Surowcami Mineralnymi i Energią PAN, Kraków.

Jędrzejec E. 1986: Properties of the r Parameter of the Budryk-Knothe Theory in the Light of Transivity of its Description of a Subsiding Trough. Prace Komisji Górniczo-Geodezyjnej PAN, Geodezja 32, Kraków.

Knothe S. 1953: Równanie profilu ostatecznie wykształconej niecki osiadania. Archiwum Górnictwa i Hutnictwa, t. 1, z. 1, Warszawa.

Knothe S. 1984: Prognozowanie wpływów eksploatacji górniczej. Wyd. "Śląsk", Katowice.

Kowalski A. 1984: Określenie zmienności parametru promienia zasięgu wpływów głównych w górotworze r_z teorii Budryka-Knothego na podstawie badań geodezyjnych przemieszczeń pionowych górotworu. Główny Instytut Górnictwa, Katowice (praca doktorska).

Kowalski A. 1985: Zmienność parametru zasięgu wpływów głównych w górotworze. Ochrona Terenów Górniczych, nr 72/2, Katowice.

Kwinta A. 2009: Transitivity Postulate Effect on Function of Influences Range Radius in Knothe Theory. Archives of Mining Sciences, Vol. 54, Issue I, Kraków.

Kwinta A. 2012: Prediction of Strain in a Shaft Caused by Underground Mining. International Journal of Rock Mechanics & Mining Sciences, Vol. 55.

Litwiniszyn J. 1962: O niektórych liniowych i nieliniowych modelach niecki osiadania w górotworze sypkim. Przegląd Górniczy, t. XVIII, Nr 5, Katowice.

Niedojadlo Z. 2008: Problematyka eksploatacji złoża miedzi z filarów ochronnych szybów w warunkach LGOM. Uczelniane Wydawnictwa Naukowo-Dydaktyczne AGH, Kraków.

Majcherczyk T., Małkowski P., Lubryka M. 2003: Naruszenie eksploatacją górniczą filarów ochronnych szybów na przykładzie KWK "Jas-Mos". Mat. Konf. Szkoły Eksploatacji Podziemnej, Kraków.

Majcherczyk T., Niedbalski Z., Wałach D. 2013: Variations in Mechanical Parameters of Rock Mass Affecting Shaft Lining. Archives of Mining Sciences, Volume 58, Issue III, Kraków.

Piwowarski W. Dżegniuk B. Niedojadło Z. 1995: Współczesne teorie ruchów górotworu i ich zastosowania. Wyd. AGH, Kraków.

Sroka A. Schober F. Sroka T. 1987: Ogólne zależności między wybraną objętością pustki poeksploatacyjnej a objętością niecki osiadania z uwzględnieniem funkcji czasu. OTG 79/1, Katowice.

International Mining Forum 2015, Kicki et. al. (eds) © 2015 Taylor & Francis Group, London, UK. ISBN 978-1-138-02820-3

State-of-the-Art in Blind Shaft Drilling for Shaft Sinking in the Coal-Mining Industry in China

Song Haiqing
Hohai University, Nanjing City
Anhui University of Science and Technology, Huainan City

Yao Zhishu
Anhui University of Science and Technology, Huainan City

Cai Haibing
Anhui University of Science and Technology, Huainan City

ABSTRACT: Blind shaft drilling method and artificial ground freezing method are the most used and reliable techniques to sink shafts in deep alluvium for coal-mining industry in China. The paper covers the development of the blind shaft drilling for vertical shaft sinking since its first use in 1968. First of all, a statistical data about shaft liner structures, drilling depth, liner diameter, liner thickness and geological conditions are given for the major and typical vertical shafts completed between 1968 to the present date. The significant advancements achieved in the procedures, such as drilling system, lining system and floating and sinking process, are also discussed in detail. A comparison of costs, construction rate, liner structure and risks is also made between the two shaft sinking techniques based on one recently completed project. Finally, an analysis of future prospects for blind shaft drilling is made for vertical shaft sinking in Chinese coal-mining industry.

1. INTRODUCTION

In underground coal mining industry, vertical shafts have to be constructed to get through water bearing soils down to the hard rock and coal seams. Special methods are always applied to sink shafts through thick and loose water bearing soils, blind shaft drilling method and ground freezing method are the recognized, reliable, safe and efficient methods for vertical shaft sinking around the world. Although the paper will review the application and advancements of blind shaft drilling for shaft sinking in Chinese coal-mining industry. But it still should be emphasized that ground freezing method also has been playing a great important role in the construction of shaft from its first application in 1955. According to the information obtained, In China in 1955, freezing technique was exactly introduced from Poland and used for the first time to successfully sink the ventilation shaft for Linxi Coal mine in Kailuan coalfield located in Hebei Province. From that time, more than 900 shafts have been sunk during the last six decades for Chinese coal-mining industry. During the period, numerous advances have been made in the use of freezing method for shaft sinking, such as multi-row freeze pipes application, frozen wall temperature control, more highly mechanized excavation, deeper freezing depth (the maximum depth is 950 m presently [1]), and larger inner diameter of shaft lining(the maximum diameter is 9 m) ,etc. So ground freezing method is approved to be a mature and competitive technique for shaft sinking in China.

Blind shaft drilling method, which is completely different from freezing method, offers several advantages over it. The process is highly automated. Using only a three- or four-man crew, all of the work is performed on the surface. No one enters the shaft during development. Depending on the geological conditions encountered, the drilling process normally advances at much quicker rates than a conventional shaft sinking operation. Usually, the process of shaft sinking with blind shaft drilling is complicated, since its first use in the 1960s, significant advancements have been achieved in the procedures of drilling system, lining system, segments floating and sinking process, and so on. During the past decades, many design institutes, constructors, research institutes and universities all contribute a lot for the progress of shaft drilling technique.

2. HISTORY OF BLIND SHAFT DRILLING IN CHINA

In 1958, China sent a research team to Soviet to investigate and study the shaft drilling technique. In 1966, drilling method was first introduced into China and successfully applied in the construction of Pansan ventilation shaft located in Anhui Province. During the past decades, over one hundred vertical shafts were completed with the drilling technique. Definitely, many relative aspects have achieved great progress, such as drilling equipment for large-hole drilling, deeper drilling depth, drilling fluid, and especially about permanent lining structure for drilled shaft. According to the data obtained [2], the total depth of drilled shafts is over 22 km already, the maximum of drilling diameter is 10.8 m, the maximum inner diameter of shaft is 8.3 m, and the maximum drilling depth for one single shaft is 660 m. Table 1 shows some details of wells drilled deeper than 400 m.

Table 1. Characteristics of completed shafts in China during the last decade [2]

Name of shaft	Depth of overburden [m]	Drilling depth [m]	Diameter of the shaft [m]		Date of completion	Location
			Drilling diameter	Inner diameter of shaft		
Longgu main shaft	546.48	581.3	8.7	5.7	05.2004	Shandong Province
Longgu ventilation shaft	546.48	578.3	9.0	6.0	05.2007	Shandong Province
Yuncheng ventilation shaft	532.29	653.0	9.0	6.0	05.2007	Shandong Province
Banji ventilation shaft	583.81	641.5	9.8	6.5	08.2007	Anhui Province
Banji main shaft	584.1	640.5	9.5	6.2	11.2007	Anhui Province
Banji auxiliary shaft	580.93	642.6	10.8	7.3	11.2007	Anhui Province
Zhangji ventilation shaft 1	401.2	458.0	10.8	8.3	05.2008	Anhui Province
Zhangji ventilation shaft 2	401.0	440.0	9.6	7.2	09.2007	Anhui Province
Zhujixi waste shaft	470.0	545.0	7.7	5.2	02.2010	Anhui Province
Xinghu central ventilation shaft	405.76	472.0	9.6	7.0	06.2012	Anhui Province

3. DRILLING SYSTEM

3.1. *Drilling equipment*

The first rotary shaft drilling rig of ZZS-1 type was developed from oil well drilling equipment during the 1960s. As mentioned before, this machine was first used in 1968 to successfully sink a shaft in Anhui Province. The drilling machine is capable of drilling shafts to a diameter of 4.3 m and to a depth of 150 m. After decades of development, the most newly developed specialized shaft drilling rig (Fig. 1) for blind shaft drilling is capable of drilling shafts to a diameter of 13.0 m and to depths in excess of a 1000 m. That is really an immense improvement from its humble beginnings.

There are three distinct periods for the development of drilling machine:
– Application of oil well drilling machines with specialized designed auxiliary equipment (1963––1972).
– Application of self-developed specialized drilling machines (1973–1983).
– Period of steady improvement for drilling technique (1983 to the present).

Table 2 lists the major drilling rigs used for blind shaft drilling during the last decades in China. Some details of drilling rigs' major performances are also listed in the table.

Table 2. Major performances of drilling rigs used in Chinese
coal mining industry and statistics of completed shafts [2]

Development period	Drilling rigs	Major performances				Number of completed shafts
		Depth of drilling [m]	Diameter of drilling [m]	Rotary torque [kN·m]	Hook lifting power [kN]	
1963-1972	ZZS-1	150	4.3	39.2	1300	8
	MZ-I	150	5.0	39.2	1300	4
	MZ-II	150	5.0	39.2	1300	2
	YZ-1	150	5.0	39.2	1300	4
1973-1983	ND-1	500	7.4	196	3200	8
	BZ-1	250	6.5	117.6	1400	1
	SZ-9/700	700	9.0	294.0	3000	10
	AS-9/500	500	9.0	294.0	3000	9
	L40/800*	600	8.0	411.6	4000	6
1983 to the present	SZ-9/700G	700	10.0	400	3000	1
	AS-9/500G	800	11.0	400	3850	2
	L40/1000	800	7.0	411.6	4000	3
	AS-12/800	800	12.0	500	6500	2
	AD120/900	900	12.0	600	7000	2
	AD130/1000	1000	13.0	600	8000	2

*L40/800 and L40/1000 type rigs were imported from WIRTH Company (Germany).

Figure 1. AD130/1000 type shaft drilling rig. Hydraulic drived rotary rig (left), drill bit (right)

Figure 2. AS-9/500 type shaft drilling rig   Figure 3. L40/1000 type shaft drilling rigs

3.2. *Drilling fluids*

The key to making the rotary drilling system work is the ability to circulate a fluid continuously down through the drill pipe, out through the bit nozzles and back to the surface.

3.2.1. *Function of drilling fluids*

Eight principal functions of a drilling mud are: 1) to carry cuttings from the bottom of the hole and carry them to the surface; 2) to cool the drill bit and lubricate its teeth; 3) to stabilize the wall of the hole with an impermeable cake to prevent it from caving-in; 4) to control formation pressure; 5) to hold cuttings in suspension when circulation is interrupted; 6) to release cuttings at the surface; 7) to support part of the weight of the drill pipe and casing in the hole; 8) to reduce to a minimum any adverse effects on the formation adjacent to the borehole.

3.2.2. *Composition of drilling fluids*

All drilling fluids, especially drilling mud, can have a wide range of chemical and physical properties. These properties are specifically designed for drilling conditions and the special problems that must be handled in drilling a well.

Three types of drilling mud are in common use:
1) Water base mud,
2) Oil base mud,
3) Emulsion mud.

For blind shaft drilling in China, the water base mud is the most used and preferred for blind shaft drilling. Water base mud consists of four components: (1) liquid water (which is the continuous phase and is used to provide initial viscosity); (2) reactive fractions to provide further viscosity and yield point; (3) inert factions to provide the required mud weight; and (4) chemical additives to control mud properties.

Because of the complicated geological conditions and drilling process, there is always no fixed mud mixture for drilling, so the chemical and physical properties of drilling muds are always changing with the geological conditions and drilling process. Table 3 give some details of chemical and physical properties of drilling muds used for three shafts drilling in Banji Coalmine located in Anhui Province, during the construction in 2009.

Table 3. Properties of drilling mud for Banji shafts

Construction process	Specific gravity [g/cm³]	Viscosity [s]	Water loss [ml/30 min]	Thickness of mud cake [mm]	PH value	Sand content [%]
Drilling	1.18 ~ 1.26	18 ~ 26	≤ 22	≤ 3.5	7 ~ 8	≤ 3
Floating and Sinking	1.18 ~ 1.22	20 ~ 22	≤ 15 ~ 18	≤ 1.5 ~ 1.8	7 ~ 8	≤ 0.5 ~ 1

4. SHAFT LINER SYSTEM

Just like drilling system, shaft liner system is also really complicated. The system usually includes selection and design of shaft lining structure, construction of shaft segments, shaft floating and sinking into the wellbore, backfilling and consolidation of shaft, etc.

4.1. *Types of shaft lining structure*

Considering the cost of construction material, concrete is the major material used in the shaft lining structure. The mostly used two types of structures are: (1) reinforced concrete shaft lining; (2) composite shaft lining structure of concrete and double-skin (or single-skin) steel cylinders. Other types of structure such as cast-iron tubbing lining and steel lining, which are popular in Germany, Russia, Ukraine and America, etc. These kinds of structures are not preferred and never used in China for the cost and complicated construction technique.

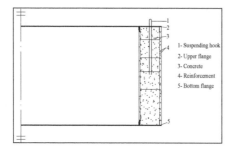

Figure 5. Reinforced concrete shaft lining

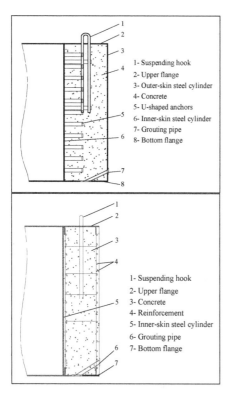

Figure 6. Composite shaft lining structure of concrete
and single-skin (double-skin) steel cylinder(s)

4.2. *Design of lining structure*

Linings for shafts are designed on the assumption that the load is evenly distributed around the structure and increases with depth. According to the equation proposed by design handbook [3], calculation of reinforced concrete shaft lining structure is based on the elastic analysis of single thick walled cylinder, only loaded by external pressure P_o as seen Figure 7 below. It has inner radius r_i and outer radius r_o. Calculation of the composite of double-skin shaft lining structure can be regarded as multi-layer composite cylinders of concrete and steel. The structure is a system of three cylinders, also loaded by external pressure P_o as seen Figure 8 below.

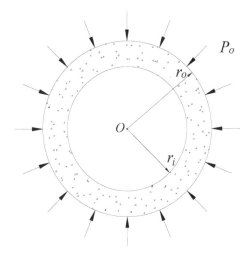

Figure 7. Thick-Walled Cylinder of Reinforced Concrete Shaft Lining

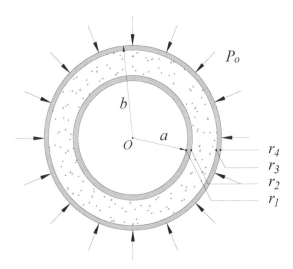

Figure 8. Multi-Layer Bonded Cylinders of Composite Shaft Lining

Wall thickness determination

(1) Reinforced concrete shaft lining structure
In order to determine wall thickness of reinforced shaft lining as shown in Figure 7 above, the stresses throughout the cylinder should be determined firstly. According to the linear elastic solution of this problem, which is attributed to the French mathematician Gabriel Lame' in 1883, the radial and hoop stresses at any radial location r, are given by the following formulas:

$$\sigma_r = \frac{-P_o r_o^2 + P_o(r_i^2 r_o^2 / r^2)}{r_o^2 - r_i^2} \tag{1a}$$

$$\sigma_\theta = \frac{-P_o r_o^2 - P_o(r_i^2 r_o^2 / r^2)}{r_o^2 - r_i^2} \tag{1b}$$

where: σ_r = radial stress, σ_θ = hoop stress, r = radius, r_i = inner radius, r_o = outer radius, P_o = external pressure.

So, as shown in Figure 9 below, stress on the inner surface $\sigma_{\theta i}$, is the maximum. And:

$$\sigma_{\theta i} = -\frac{2r_o^2 P_o}{r_o^2 - r_i^2} \tag{2}$$

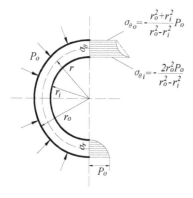

Figure 9. Stresses distribution analysis of shaft lining cylinder

Then, according to the Design code of reinforced concrete structure, the actual load (external earth pressure) should be multiplied by load safety factor v_k, to obtain the ultimate design load, and 1.4 is the minimum dead load safety factor. So:

$$|\sigma_{\theta i}| = \frac{2v_k r_o^2 P_o}{r_o^2 - r_i^2} \leq f_c \tag{3}$$

After deduction, the wall thickness:

$$h = r_i \left(\sqrt{\frac{f_c}{f_c - 2v_k P_o}} - 1 \right)$$ (4)

(2) Composite shaft lining structure of concrete and double-skin steel cylinders
As shown in Figure 2 above, the composite shaft lining structure is composed by inner and outer thin-walled cylinders and the thick-walled concrete cylinder between them. The outer radial deflection of any cylinder must equal the inner radial deflection of the cylinder that is bonded to its outer surface. This statement leads to the equation:

$$u_b^i = u_a^{i+1}, i = 1, N-1$$ (5)

where: u_a = inner radial deflection, u_b = outer radial deflection.

Based on Equation (5), the unknown interface pressures (P_i) between the cylinders can be yielded, and then they can be substituted back into the Lame's equations (Equation (1)) to determine the stresses throughout each cylinder at any arbitrary radius. Finally, the thickness of middle concrete cylinder can be determined after the same analysis as reinforced shaft lining structures.

4.3. Physical model test on shaft lining structure

Since the mid-1980s, practitioners encounter a lot of problems with design and construction of shaft lining. All of these problems can be attributed to: depth increasing of exploitation, growing technical demands and increasingly difficult geological and mining conditions, which have forced practitioners to undertake research into new types of lining structure, new techniques and new building materials.

For the contribution of blind shaft drilling technique in China, Anhui University of Science and Technology carried out a series of physical model tests about the mechanical properties of reinforced concrete shaft lining and composite shaft lining under horizontal loading since the mid-1980s. And the university has always been endeavoured to introduce the application of new building materials, such as pumpable high-strength and high-performance concrete (C60-C100 class) and steel fibre reinforced high-strength concrete (to increase the ductility of concrete). Figures 10, 11, 12 show some details about the model tests on the reinforced concrete shaft lining and composite shaft lining of concrete and double-skin steel cylinders. All of the models are placed in the self-developed hydraulic loading device for testing. For concrete shaft lining, the results of studies have shown that mechanical properties of individual segments are indirectly affected by compressive strength of concrete and the ratio λ (characterizing the relationship between the thickness of the casing wall and its inner radius). Reinforcing bars have a negligible effect on the extent of horizontal loads acquired by the casing. The value of maximum rupture load ranged from 21.5 to 31.0 MPa, depending on the class of concrete used (C65/75) and the dimensions of the model [4]. And for composite structure, as the results shown in [5], depending on the dimensions of the composite shaft lining structure models, class of concrete used (C65/75) and thickness of inner and outer steel cylinder, the maximum rupture load ranged from 29.5 MPa to 41.0 MPa. It is obviously shown that under the confined action steel cylinders, compressive strength of concrete between them improves greatly, which lead to the increase of ultimate bearing capacity for composite shaft lining structure compared to reinforced concrete structure.

a) Reinforcing steel bar mats installation

b) Concrete placing

c) Strain gauges attached onto the outer and inner surfaces

Figure 10. Preparation of reinforced concrete shaft lining model

a) U-shaped anchor plates welded onto the inner steel cylinder

b) Model before test

Figure 11. Preparation of composite shaft lining
of concrete and double-skin steel cylinders

Figure 12. Independently-developed hydraulic loading device

4.4. *Construction*

Shaft segments are fabricated on surface before sinking. Segments have to be cured carefully according to the ambient temperature because of the inside high temperature caused by hydration heat of mass concrete during the hardening period. Fabrication of shaft segments is a complicated process according the following drawings [6] (Fig. 13, 15 and 16) of Zhangji ventilation shaft provided by the design institute.

4.4.1. *Construction of reinforced concrete segment*

Figure 13 shows a segment with a height of 3 m and a thickness of 0.8 m and 8.3 m in diameter. This element has two steel flanges: the upper (I) and lower (II), which are connected with the reinforcement (①-⑤). The total mass of the segment to 183.6 tonnes, of which 15.6 tons are part of the steel (Tab. 4), and the rest is concrete (nearly 70 m³).

Upper flange (I), a steel ring with a width of 800 and a thickness of 20 mm equipped with 32 evenly spaced rectangular openings. Half of them allows for concrete placing. For the remaining 16 are allowed for hanging hooks (⑥) for suspending the element. Perpendicular to the surface of the flange are welded steel plates (A, B), which are the elements for fastening vertical reinforcement (②,④) through welding. So after concrete has hardened completely, the flanges will be tightly "glued" with concrete as a structure.

Bottom flange (II) also has elements of A and B, which are steel plates with the same size. The flange is provided with four holes for cement-based grout injecting into the gap between the bottom and upper flanges. Grout is injected under pressure through the embedded pipe (C).

The parameters of individual segment are summarized in Table 4, the final view of the upper flange of the component is shown in Figure 14.

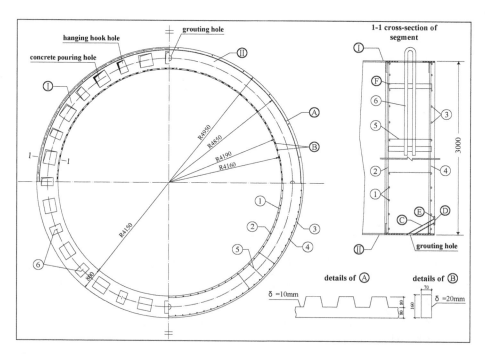

Figure 13. The structure of reinforced concrete shaft lining

Table 4. Details list of steel elements

Symbol	Diameter	Amount	Weight	Total weight		Symbol	Amount	Weight	Total weight
-	[mm]	[PCs.]	[kg]	[kg]		-	[PCs.]	[kg]	[kg]
1	28	17	128,0	2176,0		I	1	2786,9	2786,9
2	25	96	11,6	1113,6		II	1	3590,7	3590,7
3	28	17	148,7	2527,9		A	1x2	287,4	574,8
4	25	123	11,6	1426,8		B	96×2	1,8	345,6
5	10	97	0,6	58,2		C	4	4,9	19,6
6	50	16	53,5	856,0		D	4	1,1	4,4
			Σ	8158,5		E	4	1,4	5,6
				↓		F	32	4,4	140,8
				15626,9	←			Σ	7468,4

Figure 14. The reinforced concrete shaft lining and upper flange plate

4.4.2. *Construction of steel-concrete segment*

The steel-concrete segments (Fig. 15) are also connected by welding the bottom and upper flanges together (I and II), the same as reinforced concrete segment. The inner and outer skins are made of steel sheets with a thickness 20 mm (④) and 16 mm (⑤) respectively. U-shaped elements (①)

are welded onto the inner steel cylinder (④) for anchoring with concrete. During the transport segment is suspended by hooks (2) embedded into the segment via welded connectors (⑥). Cement-based grout is also injected into the gap of flanges through a pipe (③). Figure 16 shows the distribution of U-shaped anchors on the surface of inner steel cylinder.

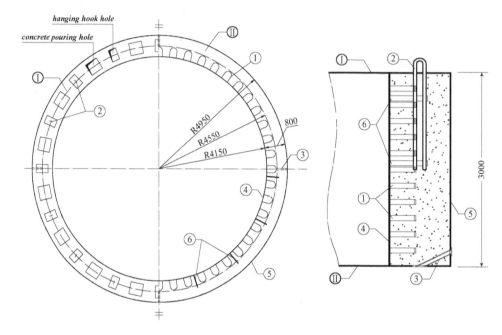

Figure 15. The composite of double-skin shaft lining structure

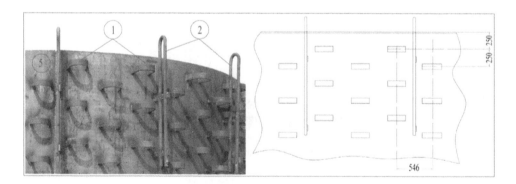

Figure 16. Distribution of the U-shaped anchor plates

4.4.3. *Construction of a reinforced concrete bottom*

Designing the structure of shaft bottom takes into account the stress caused by the formation water and blasting. The bottom is a strong structure made up of elements similar to those used in the previously described segments. The spherical shape (Fig. 17) ensures the stress distribution.

Figure 17. Reinforced concrete bottom of shaft lining

4.5. *Backfilling and Consolidation of Shaft Lining*

After the shaft lining cylinder is lowered and floating into the target depth, it is important to backfill the annulus between shaft lining and surrounding ground. As shown in Figure 11, during the backfilling process, it is necessary to ensure all drilling fluids in the annulus is expelled and replaced by backfilling material to completely fill the space. The reasons for backfilling are:
- Stabilizes the liner during construction.
- Puts the liner in full contact with the surrounding ground to allow load transfer.
- Helps reduce permeability of the final liner system.
- Reduces groundwater flow around (behind) the final liner.
- Gives some added corrosion protection to the final liner.

Backfilling material
Basically, cement-based grout (with specific gravity of 1.6~1.8) and crushed limestone (with a maximum size of 40 mm) are the mostly used backfilling material during the past decades' backfilling construction practice. As shown in Figure 19 below, cement-based grout is the only choice to backfill the annulus behind outer surface of composite shaft lining and bottom, which will ensure the tightness of backfilling body and effectively prevent outer steel cylinder from corrosion of groundwater. As for backfilling behind RC shaft lining, cement-based grout and crushed limestone are used alternately (shown in Figure 20).

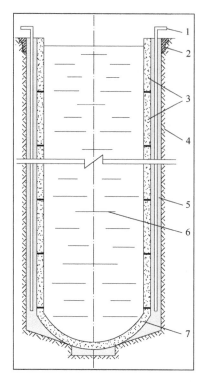

Figure 18. Schematic of annular backfilling between shaft lining and surrounding ground. 1 – Vertical grout pipes; 2 – Shaft collar; 3 – Shaft lining segments; 4 – Surrounding ground; 5 – Drilling fluid; 6 – Inpouring water; 7 – Shaft bottom

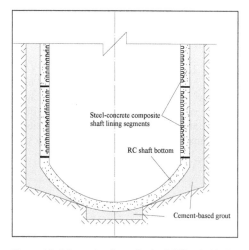

Figure 19. Schematic of annular backfilling behind composite shaft lining and bottom

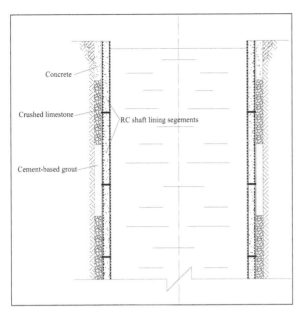

Figure 20. Schematic of annular backfilling behind RC shaft lining

5. ECONOMIC ANALYSIS BETWEEN FREEZING METHOD AND DRILLING METHOD

Besides the technical issues, economic analysis is an important factor affecting shaft sinking method choosing between freezing technique and drilling technique. Tables 5, 6, 7 give the details for comparison of project amount, construction time and cost input between the two techniques during the feasibility study for construction of Banji Coalmine mentioned above.

Table 5. Project amount comparison between freezing method and drilling method for Banji Coalmine

Sinking method	Main shaft		Auxiliary shaft		Ventilation shaft	
Subjects	Freezing method	Drilling method	Freezing method	Drilling method	Freezing method	Drilling method
Inner diameter of shaft [m]	5.8	5.8	7.0	7.0	6.5	6.5
Drilling depth [m]	640.0	640.0	625.0	625.0	640.0	640.0
Excavation volume of soil [m³]	46448	37158	61393	49063	56678	45342
Concrete [m³]	28610	10205	37352	13323	35919	12812
Reinforcement [t]	1562	558	2039	727	1961	699
Volume of Backfilling [m³]		8125		9327		9048
Steel plate [t]		1266		1129		1353

Table 6. Construction time comparison between freezing method and drilling method for Banji Coalmine

Sinking method / Subjects		Main shaft		Auxiliary shaft		Ventilation shaft	
		Freezing method	Drilling method	Freezing method	Freezing method	Drilling method	Freezing method
Inner diameter of shaft [m]		5.8	5.8	7.0	7.0	6.5	6.5
Drilling depth [m]		640.0	640.0	625.0	625.0	640.0	640.0
Construction time [months]	Preparation period	7	3	8	3.5	8	3.5
	Drilling and sinking period	15	18	17	20	17	19.5
	Total	22	21	25	23.5	25	23
Difference		21-22 = -1		23.5-25 = -1.5		23-25 = -2	

Table 7. Shaft sinking cost comparison between freezing method and drilling method for Banji Coalmine

Sinking method / Subjects		Main shaft		Auxiliary shaft		Ventilation shaft	
		Freezing method	Drilling method	Freezing method	Freezing method	Drilling method	Freezing method
Inner diameter of shaft [m]		5.8	5.8	7.0	7.0	6.5	6.5
Drilling depth [m]		640.0	640.0	625.0	625.0	640.0	640.0
Cost [thousands RMB]	Cost per meter	17.176	11.715	20.73	14.118	19.249	13.11
	Total cost	10992.64	7497.6	12956.25	8823.75	12319.36	8390.4
Cost difference [thousands RMB]		3495.04		4132.5		3928.96	
Percent of increased cost for freezing method [%]		31.8		31.9		31.9	

According to the data shown in above tables, there is little difference for construction between two sinking methods. But the cost for freezing method increase about 30% more than drilling method.

6. FUTURE PROSPECT FOR DRILLING METHOD IN SHAFT SINKING

During the past decades, the drilling method has been progressed in each aspect. But there are still some problems yet to be improved, such as: 1) improving the structure of drill bit to strengthen cutting capacity; 2) developing new type of high-strength shaft lining structure to reduce the thickness and weight of segments; 3) developing and manufacturing more powerful hydraulic rotary drilling rigs to enhance drilling rate. So blind shaft drilling method will be a more reliable, efficiently and cost-effective technique in the future for Chinese coal-mining industry.

ACKNOWLEDGEMENTS

This work was supported by the Foundation for Outstanding Young Talents in Higher Education Institutions, Anhui Province (Grant No. 2011SQRL044), and also supported by Anhui Province Natural Science Research Project (Grant No. KJ2012A094).

REFERENCES

[1] Wang J.P., Li C.Z., Xu S.R. et al. 2011: New Development of Ground Freezing Method (in Chinese). Journal of Mine Construction Technology, 2011, 32(1/2): 39–41.

[2] Zhang Y.C., Shi J.S., Wang Z.J. et al. 2010: Construction Handbook of Mine Shaft Drilling (in Chinese). China Coal Industry Publishing House, Beijing.

[3] Zhang R.L., He G.W., Li D. 2003: Design Handbook of Mining Engineering (in Chinese). China Coal Industry Publishing House, Beijing.

[4] Yao Z.S., Chang H., Rong C.X. 2006: Research on Stress and Strength of High Strength Reinforced Concrete Drilling Shaft Lining in Thick top Soils (in Chinese). Journal of China University of Mining & Technology, 2006.

[5] Yao Z.S., Chang H., Rong C.X. 2007: Experimental Study on Drilled Shaft Lining of High Performance Concrete and Double Steel Cylinder in Super-Deep Alluvium (in Chinese). Chinese Journal of Rock Mechanics and Engineering, 2007.

[6] Hefei Design Research Institute for Coal Industry, Anhui Province, 2006: Designed Drawings of Shaft Lining Structures of Zhangji Ventilation Shaft Located in Anhui Province.

International Mining Forum 2015, Kicki et. al. (eds) © 2015 Taylor & Francis Group, London, UK. ISBN 978-1-138-02820-3

Vertical Shaft Sinking for Underground Transportation Purposes in Slovenia

Marko Ranzinger
RGP d.o.o., rudarski gradbeni programi, Rudarska 6, 3320 Velenje, Slovenia
e-mail: Marko.Ranzinger@rlv.si

Marjan Hudej
RGP d.o.o., rudarski gradbeni programi, Rudarska 6, 3320 Velenje, Slovenia
e-mail: Marjan.Hudej@rlv.si

ABSTRACT: Vertical shaft sinking for underground transportation purposes is a very demanding technological process in mining and geotechnology which requires specific and specially designed technological equipment. At the same time it is a technological process that humanity deals with from its beginning, since mining is one of the oldest industries in the world.

The aim of the article "Vertical shaft sinking for underground transportation purposes in Slovenia" is to show the last two shafts sinking in Slovenia in low-strength rocks.

The first shaft sinking was for Holding of Slovenian Power Plants who invests in construction of the Pumped storage hydropower plant in Slovenia. The location of pumped storage hydro power plant is in western part of Slovenia. A lot of parts are underground. The project finished in March 2010. There is also a Vertical pressure shaft with 190 meters of depth and 4 meters of diameter for water transportation to the facility Pumped storage hydropower plant Avče. In a paper work is present all underground excavation works with support measures and a problems in this vertical pressure shaft which we had during the work.

The second shaft sinking is still in the progress. Velenje Coal Mine has made up the decision for the sinking of a coal production shaft, depth of 505 meters. In future the complete coal production (4 million tons per year) should be hoisted to surface instead of an inclined shaft by using a belt conveyor. The new shaft is foreseen for hoisting coal only and not for man riding or material-transport. We started with the preparation works in 2011 and with a shaft sinking in 2012. The aims of the paper is to show the preparatory work before the start of the actual work, describes the equipment required for the vertical shaft sinking, it shows the primary and secondary support measurements and monitoring of mining work at such facility.

With the paper we would like to represent the mining field of shaft sinking in Slovenia in last years.

KEYWORDS: Vertical shafts for underground transportation purposes, shaft sinking

1. VERTICAL PRESSURE SHAFT FOR HYDROPOWER PLANT AVČE – BACKGROUND

The location of pumped storage hydro power plant Avče (Avče PSP) is in western part of Slovenia. The powerhouse is situated on the left bank of the river Soča, downstream from the village of

Avče. The existing reservoir of the Plave hydropower plant, Ajba serve as the lower storage reservoir for pumped storage hydro power plant Avče. The upper storage reservoir is situated in a natural depression in the vicinity of the Kanalski vrh settlement. The headrace tunnel and the penstock connect the upper storage reservoir to the powerhouse (Fig. 1).

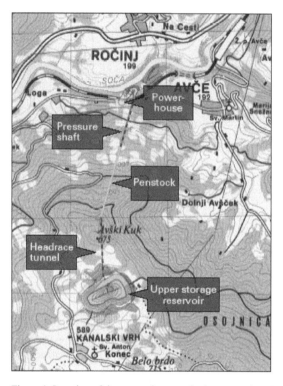

Figure 1. Location of the pumped storage hydropower plant Avče

Basic characteristics of the hydropower Avče:
- Installed capacity in turbine operation 185 MW.
- Annual average energy prod. approx. 426 GWh.
- Power plant utilization rate 0.77.
  The other objects of pumped storage hydropower plant Avče are:
- Upper storage reservoir
  Maximum headwater level 625 m asl.
  Minimum headwater level 597 m asl.
  Effective storage capacity 2.17 million m³.
- Lower storage reservoir (Ajba)
  Maximum tail water level, continuous operation 106.0 m asl.
  Minimum tail Water level, continuous operation 104.5 m asl.
  Useful daily reservoir volume 0.42 million m³.
- Penstock
  Inner diameter 3.1–3.3 m.
  Total length 862 m.

1.1. *Penstock underground alternative*

After several months of geological investigations, the open-air penstock had to be rerouted underground between T6 and T11. Chose alternative foresees a 190 meters high vertical pressure shaft followed by a 385 meters long sub-horizontal pressure tunnel. In order to excavate we made parallel the vertical shaft and the excavation of the horizontal pressure tunnel.

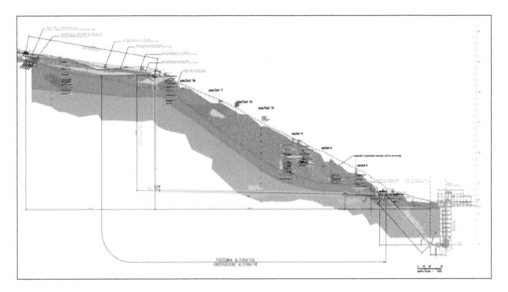

Figure 2. Profile of the underground penstock with a vertical pressure shaft (190 m) and horizontal tunnel (385 m)

1.2. *Excavation works – pre-drilled pilot hole*

We made a conventional pilot drilling with a three-cone pilot bit used 444,15 mm in the first step and 650 mm when we re-drill the hole. For a drilling we used a drilling machine FRASTE FS 400 MAN and a technology of rotation drilling with a drilling mud. Time table of the drilling is on the Table 1.

Table 1. Time table of the drilling

	Drilling with a three-cone bit diameter 444,15 mm	Drilling with a three-cone bit diameter 650 mm	Air lift
Time	12.11.–21.12.2008	7.1.–28.2.2008	28.2.–14.3.2008

Figure 3. The location of the shaft with a drilling machine

From the analysis of the drilling we got the diagram of loss the drilling mud. After the 38 meters of the pilot hole we had a rock with a lot of cracks and because of that we had problems with a drilling mud.

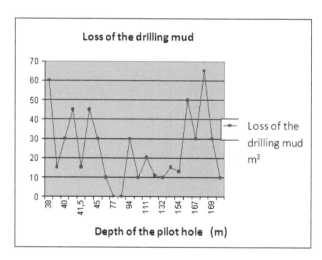

Figure 4. The loss of the drilling mud during the drilling
with the diameter of 444,15 mm

The most drilling advance with a bit of diameter 444,15 mm was 23 m/day. When we used a bit of diameter 650 mm we had the top of drilling advance 28 m/day (Fig. 5).

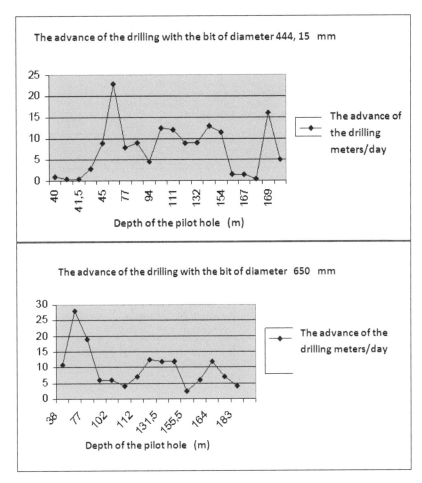

Figure 5. The advance of the drilling with the diameter of 444,15 mm and 650 mm

1.3. *Conventional vertical shaft sinking with the rock hoisted through horizontal tunnel*

Conventional shaft sinking using drill and blast techniques have been practiced for as long as underground mining taken place. Also in this case the excavation works was proceed by steps which length depends on the rock quality. Drilling and blasting of the rock in the bottom of the shaft and the mucking of the material through pilot hole was the main excavation works.

In a poor to very poor rock conditions, the length was reduced while it can increased in fair to good rock. Generally speaking the blasting length is range between 1.0 to 2,0 meters. Each step was consist of a typical excavation round with drilling, mucking, rock support. The rock support must be fully supported before proceeding with the next excavation step.

For each profile, a basic rock support has been pre-defined. It consists of a reinforced shotcrete layer with a corresponding mesh of grouted rock bolts.

The vertical pressure shaft we divided into 2 steps: (1) un upper part with a depth inferior to 40 meters and a diameter of 8,00 meters and (2) a lower part with a depth between 40 meters and 190 meters with a diameter of 4,00 to 4,40 meters.

– *Upper part (depth < 40 m)*

Three typical profiles/rock support measures have been developed in function of the expected rock conditions.

Table 2. Type of rock support as for rock quality

Rock Quality	*Rock Support*
Weathered rock	Type 1
Creeping soil	Type 2
Unstable material (completely crushed)	Type 3

For each profile, a basic rock support has been pre-defined. It consists of a reinforced shotcrete layer with a corresponding mesh of grouted rock bolts. Some additional measures with steel ribs should be applied when the rock conditions are poor to very poor. The typical profiles are as follows:

Table 3. Support measures

Profiles	*Shotcrete thickness* [cm]	*Grouted Rock Bolts*				
		Diameter [mm]	Length [m]	Number of rock bolts [pces]	Vertical spacing [m]	Mesh [m²]/ /Steel ribs TH [m]
Type 1	10–15	25	4,00	10	2,00	5,00
Type 2	20–25 with steel fabric	25	4,00	12	1,50	3,00
Type 3	20–25 with 1–2 layers steel fabric	25	4,00	Spot rock bolts	/	Steel ribs TH29

Figure 6. The machine excavation on the bottom of the shaft

Each round consists of excavation work with drilling and blasting and rock support.

– *Lower part (depth > 40 m)*

The work started in October 2007. After sinking and supporting this upper part (depth < 40 m) work starts with a lower part. After drilling of the advance hole the shaft sunk by using hand held drill hammers. Each round consists of excavation work, cleaning of shaft bottom and rock support. Excavation work is done carefully and by the application of smooth blasting methods to minimize rock disturbance.

Figure 7. In the lower part

Four typical profiles / rock support measures have been developed in function of the expected rock conditions.

Table 4. Type of rock support as for rock quality

Rock Quality	Rock Support
Very good/good rock	Type 1
Fair rock	Type 2
Poor rock	Type 3
Very poor rock	Type 4

For each profile, a basic rock support has been pre-defined. It consists of a reinforced shotcrete layer with a corresponding mesh of grouted rock bolts.

1.4. *Conclusion*

Because the geological problems, the open-air penstock had to be rerouted underground between T6 and T11. With the new shaft of 190 meters we finished the project of hydropower plant Avče. Today the hydropower plant works normally.

2. NEW PRODUCTION SHAFT »NOP II«

Premogovnik Velenje – Velenje Coal Mine is an underground coal mine with average annual production around 4 million tons of lignite coal. Existing main conveyance system, which brings whole mine production to the surface, is a system of several consecutive belt conveyors. Galleries and equipment are well maintained, system runs well and it suits its need. Despite all this, conveyance system has some weaknesses and disadvantages. It is widely spread and distant to other mine facilities; it is difficult to ensure its operational reliability; high operating costs, many kilometres of open underground galleries presents also some safety risks.

These were the main reasons for us to start thinking about new transportation system, new technical solution, which can also be more cost efficient. In 2009 we started technical and economic studies for new production shaft equipped with skips. New production shaft "NOP II" is located closer to active production panels underground, closer to stock deposits and closer to power plant on the surface. After exhaustive study and revision, at the end of 2010, we got allowance for investment and approval to start operative activities on our project.

Through technical studies we determined shaft dimensions and we chose double compartment skip with friction hoist (Koepe) as adequate hoisting system. You can find some basic technical characteristic for our new production shaft and hoisting system in Table 5.

Table 5. Technical characteristics for new production shaft and hoisting system

Shaft inner diameter	6,15 m
Shaft depth	505 m
Hoisting distance	490 m
Max. hoisting speed	12 m/s
Payload	23,0 t
Hoisting capacity/hour	964,9 t/h
Hoisting capacity/year	4.081.503 t/year

Preliminary-works have started in January 2011. In first and second phase of pre-works we prepared infrastructures for construction site, we constructed shaft collar and we sank shaft to depth of 37 m using crane at open shaft collar. In April 2012 we started with assembly and installation of hanging platforms, top cover, temporary headframe and winch hall with nine winches. Complete set of shaft sinking equipment was ready in September 2012 when we started with shaft sinking. At this moment (September 2014) we are at depth 285 meters.

2.1. *Equipment in the shaft sinking technology*

On location of new shaft detailed geologic exploration was done and we had some experience information from shafts which were sunk in the same area years ago. Therefore we knew exactly what will be geological and hydrological conditions for shaft sinking. We decided to use conventional mining shaft sinking method; using hanging platforms in the shaft, drum hoists, drum winches and temporary headframe on the surface and other auxiliary equipment for dewatering, ventilation, gas monitoring, etc. Hanging platforms headframe and other steel structures were designed by experts at Premogovnik Velenje and manufactured at our subsidiary companies.

Figure 8. Shaft site – surface

Work site – bottom of the shaft
At the bottom of the shaft we installed electrohydraulic loader which can excavate profile of the shaft and load muck in the kibbles. Loader is small and movable on tracks; it can also rotate 360° around its axis. It can be lifted from the bottom by the main drum hoist and in few separate parts it can be transported through hanging platforms up to the surface. At work site you can find signalling and communication equipment, gas monitoring equipment, dewatering pump and pipelines, pneumatic tools. Fresh air is coming from surface, pushed by ventilator down the shaft, through 800 mm ventilation tube.

Figure 9. Work site

Working platform

At maximum of ten meters above work site is positioned first hanging platform, so called working platform. Working platform has two levels. First level can be used as platform from which we can execute certain operations, but at the same time it is protection shield above work site at the bottom. Second level of working platform is construction formwork for final concrete lining.

Tensioning platform

Some 30 up to 50 m above work site is positioned second hanging platform, so called tensioning platform. Its weight is about 40 t, it has three levels, and its main purpose is tensioning of the guidance ropes. On the first level we have water container and dewatering pump and there is electric equipment for power supply to the work site. Tensioning platform is also used for installation of shaft insets including final ladder compartment. Ladder compartment is installed simultaneously as final concrete lining is done. Dewatering pipeline, pipelines for compressed air and shotcrete are installed at the same time. The rest of shaft insets, including bunton layers and guidance rails for the skips, will be installed at the end, when shaft sinking is completed. Beneath working platform and tensioning platforms we are using rope ladders as emergency escape way.

Both hanging platforms have openings to enable kibbles to pass through on their way to the bottom/surface. Kibbles are guided by guiding frame - yoke which slides along guidance ropes to the tensioning platform; beneath tensioning platform kibbles are driven unguided.

Sounding line platform

Sounding line platform is fixed level about 10 m beneath surface. There we have installed bridge platform which carries the central axis sounding line. This is the most important sounding line used for shaft cantering. Additional on this level we have installed six more sounding lines. Three of them are already used for accurate installation of main holders of ladder compartment, and the rest of them will be used later at installation of guidance rails.

Top cover

On the surface, on shaft collar, we have top cover with 1,5 m high parapet wall, which protects shaft entry, allows used air exhaust and enables all infrastructure to enter in to the shaft. Top cover is massive steel construction which also carries half of load on tensioning platform. Tail ends of guidance ropes are connected to holders of top cover with special hydraulic system, which enables us to adjust tensioning force in each rope separately. On top cover we have two openings with pneumatic flaps, entry to inspection platform and to ladder compartment.

Headframe

On surface we have 37 m high headframe with deflection sheaves for all ropes. Headframe has smaller inner frame with manger slide for muck which is used for emptying kibbles. Particular pneumatic tool in headframe also enables us to separate kibble from guiding yoke to exchange kibble. We are using three different type of kibbles for man-riding and transportation of different material.

Winch hall and winches

We are using two independent drum hoists for man-riding and transportation of material. Then we have four winches to manoeuvre tensioning platform, two winches to move working platform and one more winch for manoeuvring main energetic cable. All together nine winches, which are installed on massive foundation in winch hall. In winch hall we have a special room for electric equipment and separate control room for winch operator.

Safety signalling system and communication system
Modern safety signalling system was designed and installed by our personnel. System also provides digital and analog signals to safety circuit of hoisting winches and to visualization display on operators control panel.

For communication we are using three systems; conventional telephone line, open line communication units and portable wireless communication units. All three systems are designed to operate in explosive atmosphere (ATEX M1, Ex ia).

2.2. *Shaft sinking technology*

Electrohydraulic loader on the bottom is used to excavate profile of the shaft and load muck in the kibbles. Profile is excavated in steps of 0,75 up to 1,5 m, depending on geologic conditions.

For primary support we are using steel arches and shotcrete reinforced with steel mesh, which is installed in two layers. In difficult geologic conditions we were also using piles, anchors and some draining materials (mesh, geotextile and draining tubes).

Achieved average daily advancing at excavation and primary support is 1,8 m/day.

Every 20 up to 25 m we increase excavation cross section to make intermediate ring foundation with final inner diameter 6,15 m. Then we continue with shaft sinking and we excavate next 20 up to 25 m. While we excavate next section, we are surveying upper section (pre-installed extensometers, convergences). If there are no movements or deformations we can make final concrete lining.

For concreting we must move working platform with construction formwork for final concrete lining to last intermediate ring foundation and fix it in position for concreting. Final monolithic concrete lining is 500 mm thick and also reinforced with steel mesh; of course steel reinforcement must be installed in correct position before formwork is moved and fixed. Concrete with special characteristics is prepared in local concrete-plant and transported to shaft site. Further, down shaft, it is transported in kibbles designed for concrete transportation. One step of concreting final lining is up to 3 m high (which is determined by formwork height) and we are doing it in 24 hour cycle; so next day we move up working platform with formwork for next 3 m and execute concreting in next section. Depending on distance between intermediate ring foundations, and distance between openings for main holder for ladder compartment, sometimes we must adjust/reduce height of concreting step. Average efficacy at concreting final lining is 2,7 m/day.

Figure 10. Shaft, ladder compartment

When final concrete lining is finished, we move working platform in lower position. Then we make manoeuvre with tensioning platform and simultaneously we install final ladder compartment and all pipelines. Average efficacy at this phase is 7,5 m/day.

One shift crew is ten up to eleven workers including electrician and mechanic for basic maintenance. Work is running in three shifts, seven days a week. Average efficacy at complete shaft sinking is 0,6 m/day.

Geologic conditions are now much better and according to geologic prospection they will remain good at least for next 180 m; personnel is now more skilled; all start up troubles with organization and equipment are eliminated; considering this facts it is reasonable to expect better results in future.

2.3. Geology

As already mentioned, geology on location for new shaft was well known even before we started with the project. Despite this fact we execute additional exploration drilling. Borehole JUG – 48/09 (depth 521 m) was drilled in the profile of new shaft. Core from this borehole was used for exact geologic prospection.

From geologic prospection we knew that in first 100 m we will have to cross three packages of geologic layers which are containing water (sand, gravel). For more precise hydrogeologic survey another borehole, JUG – 49/09 (depth 175 m), was drilled in distance of 35 m from shaft centre. Water affluence was estimated in all three packages, and pressure sensors were installed in separate layers for latter monitoring. According to this hydrogeologic survey anticipation for water afflux during shaft sinking was 150 up to 320 l/min, from all three packages.

Actual water afflux at excavation in first package (depth 28–39 m) was 80 l/min at the beginning and later dropped to 40 l/min. We were prepared for dewatering and this afflux cause no problem.

In second package (depth 49–60 m) water afflux was much bigger. At depth of 52 m we had water and fine sands irruption which stopped us for several days. Water afflux was more than 200 l/min at the beginning, later reduced and stabilized at 80 l/min. Presence of gases was detected. At the time of first irruption concentration of methane – CH_4 rises up to 7 vol. %. We were pumping water and removing irrupted sand for several days before water pressure has dropped. We were constantly monitoring gas and water pressure by pressure sensors which were installed in borehole JUG – 49/09.

With experience from second package, we were very careful when we were approaching third package (depth 85–95 m). We execute some additional exploration drilling, but there was almost no water afflux from this layer. Permeability to gas and water in this package was much lower and we crossed it with no problems.

According to geologic prospection geologic conditions will remain good at least for next 180 m. Later we will have to cross several meters of more solid and hard layers. Excavation will probably be more difficult and slower, but we do not expect any significant problems.

2.4. Geotechnical surveillance

For such pretentious underground construction as mine shaft, it is very important to have adequate geotechnical surveillance. Certain technical parameters about construction itself and its influence to surrounding must be constantly monitored.

To achieve demanded accuracy at shaft sinking geodetic surveillance is very important. Beside sounding and guiding at excavation and concreting the final lining geodetic surveillance also includes ground movement surveillance on the surface around the shaft. Additionally we have three vertical borehole inclinometers, which are located around the shaft, to observe possible horizontal movements in top 100 m strata.

Geologic and hydrogeologic observation is done on daily basis.

Geotechnical profiles (extensometers, convergence profiles) are installed in primary support when needed (depending on geologic conditions).

Of course, there are several more technical parameters which are measured and monitored. For instance technical parameters about concrete quality control or technical parameters about ventilation in the shaft.

CONCLUSION

Shaft sinking project for new production shaft at Premogovnik Velenje, is running well. At the moment shaft depth is 285 m. We have successfully sunk the shaft through first 100 m, which was geologically most difficult part of whole 505 m. We are planning to finish shaft sinking till March 2015.

REFERENCES

[1] Junge M.: Technical study. P017-09/002, Production shaft – Premogovnik Velenje. STIEPF AG, 2009.
[2] Fuhrmann J., Koch M.: Technical study. O-26234, Velenje Main Shaft Hoisting Installation "NOP II". Siemag Tecberg, 2011.
[3] Čižmek D., Golob L., Lajlar B.: A New Production Shaft at Velenje Coal Mine. Premogovnik Velenje, IV. Balkanmine Congress, Ljubljana 2011.
[4] Golob L., Lajlar B., Kamenik M., Rovsnik M.: Investicijski program INVD NOP II ver. 2010/7A, Racionalizacija glavnega odvoza, Premogovnik Velenje, 2010.
[5] Lajlar B.: RP-370/2010BL, Izvedba pripravljalnih del za izdelavo jaška NOP II, Premogovnik Velenje, 2010.
[6] Lajlar B. et al.: RP-368/2010BL, Izdelava jaška NOP II, Premogovnik Velenje, 2010.
[7] SENG. 2013: ČHE Avče (interno gradivo). Nova Gorica.

International Mining Forum 2015, Kicki et. al. (eds) © 2015 Taylor & Francis Group, London, UK. ISBN 978-1-138-02820-3

Freezing Method as the Optimal Way of Rock Mass Treating in Polish Copper Mines

Sławomir Fabich
KGHM Cuprum Sp. z o.o. Centrum Badawczo-Rozwojowe

Jacek Kulicki
Przedsiębiorstwo Budowy Kopalń PeBeKa S.A.

Sławomir Świtoń
KGHM Cuprum Sp. z o.o. Centrum Badawczo-Rozwojowe

ABSTRACT: Since the very beginning of Polish copper mining, due to difficult hydro-geological conditions, special treatment of highly saturated Cenosoic rocks during shaft sinking was required. Shaft freezing has been chosen as a method to consolidate these rocks. Due to the rock mass structure and water inflow potential, shaft sinking without freezing was considered as highly dangerous. In the course of time, when the experience was evolving, rock mass freezing technology became safer and more reliable. New techniques appeared not only in the technical method of freezing, but also in the matter of controlling the process as well as its designing. Computer aided techniques allow engineers to make a detailed analysis and view of freeze wall development. In this paper, the evolution of knowledge and engineering experience taken in Polish copper mines and current knowledge level in that matter were shown.

KEYWORDS: Mine shafts, rock freezing, shaft sinking

1. INTRODUCTION

Polish copper mines are located in western Poland, and the deposit is at a depth between 600 and 1300 m. Rapid development of Polish copper mines in early 1960s required a large number of shafts to be sunk. As of today there are 28 shafts in operation, two of them are already closed, and one shaft is in the development stage. All of these shafts were developed in difficult geological and hydro-geological conditions related to highly saturated and unstable Cenozoic rocks. To guarantee the safe conditions during shaft sinking, rock mass freezing was chosen. Firstly this method was used in 1960 for sinking the L-III shaft. At the beginning the knowledge of geological conditions as well as the hydro-geological conditions was insufficient, and the technology of freezing was based only on experience taken from other countries. It was based only on technical knowledge rather than on practical experience. As a result, for the first shafts that adopted freezing technology, the scheme was unsuited to their sinking. As a consequence of developing this technique for the first time, brine was leaking to the frozen rock mass, leading to the thaw and, in extreme situations, to flood water entering into the shaft excavation. The archival documents indicate that the most spectacular shaft wall failures in the frozen zone took place in shafts: L-III on 07/28/1961, at the depth of 149.5 m (water flow through a hole drilled in the axis of the shaft in the amount of 34.2 m³/min), L-II on 17/11/1962, at the depth of 174.4 m (water supply from the side walls in the amount of 33.6 m³/min) and L-I on 13/07/1964, in the depth interval of 259.0–307 m (water supply from the side walls in the amount of 37.5 m³/min). These phenomena were an unpleasant surprise. Initially the reason for these shaft wall failures was attributed to an inadequate strength of freeze pipes and the extremely difficult geological and hydro-geological conditions. This interpretation was based only on cursory evaluations, since the lack of consistent monitoring of frozen rock mass condition did not allow for in-depth analysis of the process.

As a result of those failures, freezing technology has been developed to improve safety and efficiency during the sinking of further shafts in the Legnica-Glogow Copper Basin (LGOM). The first major change in technology of freezing was implemented in 1965, when laminar flow of brine between inner pipe and outer pipe was switched into turbulent flow. This significantly improved the amount of heat received by the surface of the freezing pipe and, consequently, the amount of heat received from the freezing rock, what was directly affecting the rate of development and temperature, and as a consequence, the strength of the freezing wall.

Simultaneously with the changes made in the process of freezing, the changes in the process of controlling the frozen rock mass were implemented, allowing the determination of the current freezing wall thickness, forecast of its development over time, and as a result to control the entire process of freezing.

2. GEOLOGY

The LGOM area mine shafts are developed in the difficult hydro-geological environment. The rock strength is also an issue. Up to the 380 m depth there are Cenozoic loose (soil) and cohesive rocks mostly saturated with water (quartz sands, gravel and clay, with intrusions of lignite – a few meters thickness). Due to the mentioned conditions, before shaft sinking process is started, the rock mass has to be frozen. Those are also the reason why cast-iron tubing lining (40–110 mm thickness) with concrete backfilling (60 cm thickness) has to be used.

Below the Cenozoic deposits up to the depth of 980 m Mesozoic rocks are deposited. These rocks are represented only by lower Triassic – Bunter (consisting Rhaetian middle and lower mottled sandstone).

Sandstones of middle mottled sandstone (Bunter) are also greatly differentiated. Their structure consists mostly of quartz sandstones with clay and heavily cleaved clayish-carbonate binder. The strength of these rocks is 17–85 MPa.

3. FREEZING RING LAYOUTS USED IN LGOM AREA

Freezing of the rock mass for the sinking of each and every shaft became a source of knowledge, and in connection with constant evolution of freezing wall control process it was a base for development of this technique up to the actual state. It is hard to say if it is perfect now, because as with every technology it is evolving, nevertheless if we base our techniques on the examples from the last shafts that were sunk in the LGOM area, we are able to say that the actual knowledge and experience level give a high amount of certainty that a freezing wall with required parameters (thickness and strength) will be developed in the expected time, without any malfunctions. If we analyse the evolution of this knowledge and experience we can characterise six main systems of freezing (Tab. 1 and Fig. 2 and 3).

Table 1. Freeze system designs used in LGOM area [5]

Parameter	Freezing system					
	I	II	III	IV	V	VI
Amount of freezing circles	1	1	2	1	1	1
Diameter of freezing circles [m]	13.0	13.0	10.5/13.0	14.0	16.0	16.0
Nominal distance between freezing pipes [m]	1.10	1.10	1.5/1.85	1.0	1.25	1.25/2.5
Number of freezing sections	3	2	1/2	1	1	1
Flow type	Laminar				turbulent	
Shaft diameter [m]	6.0	6.0	6.0	6.0	7.5	7.5
Type of annulus pipes	Steel				polyurethane	
Type of pipes connection	Threaded					welded

Freezing in three or two sections (I[st] and IInd systems) was done only in very few shafts at the very beginning of LGOM area for mine development, and it was very problematic due to a high frequency of malfunctions. The only advantage of those systems was quick time of freezing in the relatively challenging shallow layers of rock mass, allowing the shaft sinking to start after the rock freezing had started, without necessity to wait after preliminary freezing was done. The main disadvantage of those systems was that the deeper layers were unfrozen. This particular methodology, combined with long openings in the excavation wall, was the source of a series of failures. In another shafts some improvements were made as a result of reducing the size of the openings and applying a preliminary freeze, but there were also sections in the rock mass where walls deformation occurred and, as a consequence, it fractured the freeze pipes. It can be said that a number of the shaft wall failures could be prevented if adequate freeze wall control methodology was used. Unfortunately, problems with unfrozen walls were revealed only when a specific part of the rock mass was extracted, and then it was too late to prevent those failures.

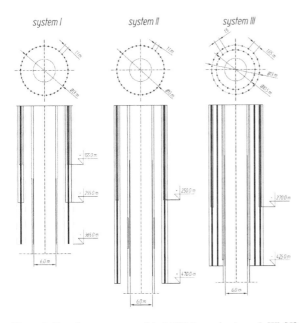

Figure 1. Freezing systems used in LGOM area (systems I–III) [5]

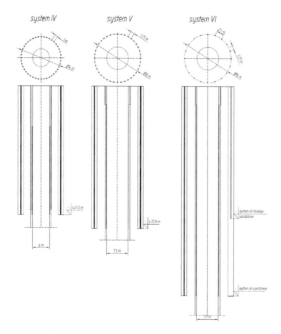

Figure 2. Freezing systems used in LGOM area (systems IV–VI) [5]

Freezing with two rings of freeze holes (IIIrd system) was only used in the P-III shaft. This system gave a very good quality of rock mass freezing. The mismatched diameter of freezing holes rings, in particular of the inner circle, was its basic disadvantage. Diameter of Φ10, 5 m resulted in uncovering freezing pipes in the openings of the shaft curb. Uncovered pipes were deformed, simultaneously destroying the initial excavation walls. In fact, only two freeze pipes of inner ring stayed intact. However the outer circle pipes remained intact.

The system applied in the P-IV, PV and P-VI shafts were a significant improvement in terms of the success of shaft sinking in the frozen zone. The above-mentioned shafts were sunk using short excavation rounds (1.5–3.0 m) and the temporary wall around the entire freezing section was installed. In the frozen sections of previously sunken shafts, this solution was used occasionally. Due to larger opening resulting from the application of temporary wall, the diameter of freeze holes ring was set at 14.0 m. For the three shafts sunk using this system, only two freeze holes were damaged (one at each shaft: P-IV and P-V). Pipes failure did not occur at the P-VI shaft.

Implementation of the Vth system took place in the second stage of the LGOM area expansion, occurring after 1970. This system was implemented simultaneously with the change of shaft sinking technology - sinking was conducted using short openings of 1.5 m in height with the final tubing lining installed. This technology is applicable today. Initially, in areas where the stresses exceeded the rock mass strength, in the weakest locations the freeze pipes were damaged. This damage was due both the deformation of the rock mass and the freeze pipes to the delayed strength gain of the primary support (setting time of concrete). The time extension of the initial freezing eliminated the rock mass deformation and alleviated these problems. In this system, the ring of freeze holes had a diameter of 16 m and the excavation wall opening diameter was 9.22–9.30 m. Also turbulent brine flow between the inner and outer pipes was used for the first time. These two last elements in conjunction with the ability to further lower the brine temperature allowed to the increase the nominal distance between the freeze holes from 1.0 to 1.1–1.25 m.

The currently used VIth system is a modification of the Vth system. It is a result of necessity for a variable depth of freezing. Therefore, this system is called deep freezing. In this system, the freeze holes are drilled from the Φ16 m diameter ring of holes, so that the every other hole is drilled to the maximum depth of the required frozen zone. Therefore, these rocks, where the thickest freezing wall is required and which have the lowest ability to be frozen, are frozen using all the holes drilled. Otherwise, the deepest and most easily frozen sandstones demand only half of the amount of holes. This system was used successfully in the recently sunk R-XI and SW-4 shafts. The gained experience concerning the operation of Vth system in conjunction with the transition to welded freeze pipe joints led to the conclusion that freeze wall using VIth system will continue to be failure-free.

4. FREEZING TECHNOLOGY EVOLUTION

Applied changes in the refrigeration plants, which are the heart of the freezing process, as well as in the collector for distributing the brine to the individual freeze hole structure, were brought about by experience acquired during the freezing of LGOM shafts. Changes in the refrigeration plants were mainly concerned with the types of applied compressors and pump productivity, resulted in a significant improvement of rock mass freezing achieved in the field. The implemented changes in the hole design, including changing the freeze pipe connections from threaded to welded, which significantly reduced the likelihood of pipe damage due to minor ground deformation. Although it is difficult to determine how well this change has reduced the possibility of pipe damage, since the nature of freeze pipe operation in the frozen rock mass is mainly the function of the freeze wall development, which was generally very good in the period after the implementation of this modification.

4.1. *Refrigeration plants*

The most important technical data concerning the refrigeration plants used in LGOM are shown in Table 2. The first freezing stations (applied in I[th]–IVth freezing systems) were equipped with ammonia single-stage compressors which allowed reduced the brine temperature to a maximum – 25°C, in winter it could be further cooled by an additional 2–3°C.

For V[th] freezing system, the freeze stations were equipped with two-stage compressors, reducing the brine temperature down to -30°C. Replacing the outdated two-stage reciprocating 8W200//2A compressors with the rotary screw SR155A compressors was a significant change in refrigeration plant design. A 41% power gain was obtained for each compressor and the amount of station refrigerant (NH_4) was reduced from 20 tonnes at R-XI shaft to 1.8 tonnes at SW-4 shaft. The amount of water to cool the compressor was reduced by 4-fold. A monitoring system of station performance was also introduced. The operator of the refrigeration plant has the ability to control processes in the equipment and to set the parameters through a wireless network remotely. In addition, all important processes such as: brine circulation within the evaporator and shaft, refrigeration plant cooling water circulation, circulation of ammonia in aggregates, control system of ammonia leakage into the atmosphere are monitored.

Figure 4. SW-4 shaft freezing plant [7]

Figure 5. Freeze plant for older freeze installations [7]

Table 2. Freeze plants used in LGOM area, major parameters

Id	Parameter	Unit of measure	Freezing system					
			I, II		II, III, IV		V	VI
1.	Period of station development	-	1960-1962		1962-1964		1970-2003	2003-now
2.	Shafts diameter	m	6.0				7.5	
3.	Depth of freezing	m	312-470		425		440-500	630-690
4.	The power of freezing stations	kW	2600		3450		3500	3900
	Thermal efficiency for two shafts		1860		2320		3490	-
	Thermal efficiency for one shaft		700		-		2790	3300
5.	Cooling device	-	single-stage				two-stage	Screw
	Compressor type	-	S4×225	3AGT	S4×225	3AGT	8W200/2A	SR155A
	Thermal efficiency	kW	122	218	122	218	349	493
	Engine power		125	320	125	320	200	355
6.	Number of compressors	pcs						
	For one shaft		6				8	8
	For two shafts		10	3	16	2	10	-
7.	Brine pump type		E1330B	OS125/3	OS200/3		E21388	100-250C
	Number of brine pumps		3	3	2		7	4
	Brine system volume	m³	190 to 280		260		300 to 340	375
8.	The average initial brine temperature at the freezing holes	°C	-20		-25		-30	-32
9.	The water consumption for cooling	m³/min	2,5		4,0		0,8	0,2

4.2. Structure of freeze hole and brine flow parameters

The freeze pipes in I–IVth freezing systems had a diameter of Φ163 or Φ141 mm with diameter of the steel inner pipe of Φ50 mm. With a brine pump flow of 3–5 m^3/min applied in the above systems (Tab. 2) it was typical to obtain a laminar flow of the brine in the space between the outer and inner pipes. As a consequence of the laminar flow, only a small amount of heat was removed by the outer surface of the freeze pipes due to the fact that the warmer and cooler zones of brine moving along parallel to the freeze hole wall path way do not mix with each other. Due to use of laminar flow in I–IVth freeze systems cold was not distributed evenly inside the freeze pipe. This phenomenon, in connection with relatively high values of initial brine temperature at the freeze holes resulted with not entirely frozen lower parts of the rock mass.

Table 3. Brine flow parameters

Brine flow parameters	Unit of measure	System I, II	System II, III, IV	System V	System VI
Rate of brine flow in the freezing plant	m^3/min	3 to 4	4 to 5	12 to 13	15
Brine flow velocity in the space between the outer and inner pipes	m/s	0,08 to 0,1	0,2	0,5-0,6	1,4
Nature of the brine flow	-	Laminar		Turbulent	

Change of the inner pipe diameter to Φ75 mm, freezing pipe diameter of Φ141 mm and the increment of the brine pumps efficiency to 12 to 13 m^3/min (Tab. 3), introduced in system V, allowed the brine to obtain a turbulent flow. This resulted with a cold flow to the pipe wall from the entire brine volume flowing across the space between the pipes. As a consequence of this change, the amount of heat taken by the surface of the freezing pipe increased almost twice.

Enlargement of the inner pipe external diameter to Φ85, and increment of brine flow rate (over 15 m^3/min), that were introduced in system number VI, allowed to the make heat transfer coefficient between freezing pipe and rock-mass higher (to about 1765 kJ/m · h · K). Currently it is 4 to 8 times higher than in the first years of the freeze method implementation in LGOM. This effort, supported by wisely chosen refrigeration plant parameters allowed for an increase in performance and productivity, and ensured the trouble-free achievement of the designed freeze wall.

Another factor contributing to the safety of the freezing wall was the change of freeze pipes connections used in the VIth system. Applied in the I–Vth systems threaded connections have been replaced with welded joints. The above-mentioned solution, in conjunction with a better frozen state of rock mass achieved in this system led to a situation where there was no failure of any of the freezing pipes in the R-XI and SW-4 shafts.

In modern freezing systems, freeze pipes are connected with refrigeration plant with two isolated pipes with built-in moisture containment and polyurethane foam insulation, located in the ground, to minimise heat losses. Each of the freeze holes is equipped with flow meters to provide independent monitoring of the brine flow inside them. In case of pipe failure there is a possibility of easy and quick identification of the failed pipe, and then exclusion of the broken hole from the brine circuit. The brine temperature is also monitored at the main inlet collector as well as at each of the freeze pipe outlets.

5. FREEZE WALL DESIGNING

The evolution of the experience gathered during shaft sinking in the LGOM area is directly correlated to the manner of freeze wall design. Due to similar geological and hydro-geological conditions resulting from the typical (with a few exceptions) structure of the Fore-Sudetic Monocline, assumptions that were placed at the start of each subsequent shaft design, and contained the sum of the experience gained during the rock mass freezing in the previous shafts. Over the years, the technology of freeze wall design was based on analytical methods of heat balance. Currently, the Finite Element Method is used for those calculations (Fig. 6). It takes into account the location of the freeze holes as well as thermal properties of the rock mass. Accordingly, the designs performed with modern numerical techniques will give more accurate results in the form of the required freezing time of the individual rock strata, rather than using the analytical methods. Application of numerical methods makes it possible to design the optimal freezing process, which is critical due to the fact that the rock mass freezing is the most expensive method for preparation of rock mass for shaft sinking.

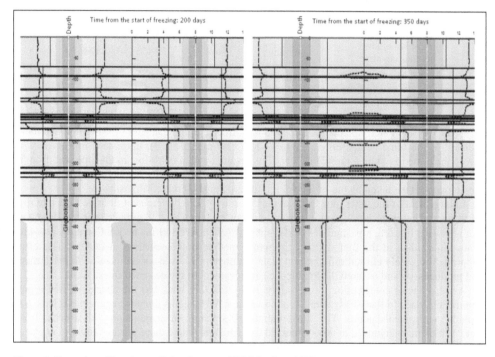

Figure 6. Examples of freezing wall development (FEM Analyses) [2]

6. FREEZE WALL CONTROL

At the beginning of rock mass freezing in the LGOM area, the only way of controlling the freezing wall was to check the difference between the brine temperature inlet and outlet. This was valuable to calculate the amount of heat that was taken up by the rock mass. The rate of freezing in the specific freeze hole was not needed anymore when the difference in the brine inlet and outlet was 1–2 °C.

This notion of control was related to the approach that deals only with freezing wall produced inside the freeze hole rings and is limited by shaft excavation wall as a load-bearing wall. The outer wall formed outside of the ring was treated only as a heat-insulating wall. This approach could not be used in the LGOM area because the required thickness of the freezing wall resulting from geological and hydro-geological conditions had to take into account its external layer. This element was included in the first shaft freezing projects, but it did not meet some basic requirements in terms of entire wall control method. In conjunction with significant freeze hole deviation, this limited the understanding of actual hole spacing.

In practice, when the Vth freezing system was used for freezing during the sinking of a shaft of a nominal diameter of 7.5 m and for greater depths of freezing than was previously used in LGOM mining area, appropriate method of freezing wall control and interpretation of their results became a priority. Control of the freeze wall temperature was carried out in:
- thermal holes located outside of the freeze ring;
- freeze holes in case of drill hole deviation;
- the wall of the shaft opening;
- rock mass behind the shaft support (tubing lining): measurements during the rock thawing phase.

In addition to the above, geo-acoustic measurements are carried out on pairs of holes, before and during the freezing process, as well as temperature measurements and brine flow rates for each freezing hole were carried out continuously in the refrigeration plants. Practice has shown that the currently used method of freezing wall controlling is suitable for monitoring its growth.

SUMMARY

In summary, the current level of knowledge derived from the experience of many generations of engineers and technicians dealing with rock mass freezing in the LGOM region, is undeniably at a level that guarantees the achievement of safe working conditions during shaft sinking. This finding is extremely important because of the increasingly difficult conditions while developing of freeze wall with required parameters in terms of static and technological considerations, primarily due to hydro-geological conditions forcing an increased freeze wall thickness. The most important actions that were taken during the recent fifty years of using this technology in the LGOM area include:
- optimal adjustment of freeze hole circle diameter and the distance between the holes for development of a freeze wall of required parameters in terms of static and technological considerations;
- increase of brine flow velocity in the space between the outer and inner pipes and increase of their geometry, in relation to the solutions used in the early period of freezing techniques implementation in the LGOM area; that allows the turbulent nature of the flow, resulting in 4–8-fold increase in the amount of heat transferred by the surface of the freezing pipe and, consequently, increase the amount of heat transferred from the freezing rock mass;
- changes in the design of freeze holes concerned the replacement of steel inner pipes for pipes made of polyethylene and replacement of the threaded connections of outer freezing pipes for welded connections. The first-mentioned solution contributed to the reduction of heat losses during the brine flow in the pipe. Otherwise the change of joints in conjunction with a good state of rock mass freezing has led to a situation that in R-XI and SW-4 shafts there was no failure of any of the freeze pipes;
- changes in the collector structure carrying brine from the freeze holes and the application of flow meters at each of the freeze holes in order to conduct the independent monitoring of the brine flow inside them, so that in case of failure of a hole there is the possibility of easy and quick identification of it and then isolation of the hole from the brine circuit;
- replacement of reciprocating compressors for screw compressors. This allows for a 41% capacity gain for each compressor, reduction of the station refrigerant (NH4) from 20–1.8 tonnes and 4-fold reduction of the amount of water intended for the compressor cooling in terms of a similar cooling station capacity;

- implementation of full monitoring program for the freeze plant for tracking and recording the most important parameters of its work;
- adaptation of measurement methods of freezing wall parameters to ensure full knowledge of its strength, geometry and localisation with regards to the shaft opening in different phases of its development and thaw after switching off the freezing system;
- implementation of modern freeze wall design methods relating to better forecasts freeze wall thickness; this is especially important in rock strata of diverse and different deterministic thermal property values.

These changes, although introduced and tested with regard to their effective on rock mass freezing for the local geological and hydro-geological conditions, can also be directly used in the worldwide mining industry. The need to share our knowledge through this article is particularly important because on the market there are limited developments in literature in terms of artificial ground freezing practices.

REFERENCES

[1] Fabich Sł., Świtoń Sł. 2010: Design of Rock Mass Freezing for GG-1 Shaft Sinking (in Polish).
[2] Fabich Sł., Świtoń Sł. 2011: GG-1 Freeze Wall Control Design (in Polish).
[3] GEO-SLOPE International 2012: TEMP/W Thermal Analysis Handbook.
[4] Ostrowski W.J.S. 1967: Design Aspects of Ground Consolidation by the Freezing Method for Shaft Sinking in Saskatchewan.
[5] Pleśniak I. 1995: Describing the Parameters and Technology of Deep Rock Freezing in Complicated Geological Conditions. Part I. Evaluation of Freeze Wall Development in LGOM Area (in Polish).
[6] Walewski J. 1965: Rules for Mines Design. Part V. Designing of Shafts (in Polish).
[7] Photographs: Archive of PeBeKa Lubin.

International Mining Forum 2015, Kicki et. al. (eds) © 2015 Taylor & Francis Group, London, UK. ISBN 978-1-138-02820-3

About the Application of Conventional and Advanced Freeze Circle Design Methods for the Ust-Jaiwa Freeze Shaft Project

Nikolai Hentrich, Juergen Franz
Deilmann-Haniel GmbH, Germany

ABSTRACT: The freeze shaft method has been used in the mining industry since the late 1800s. During the period of shaft construction, this method provides temporary ground support and water control through artificial ground freezing.

There has been significant progress in the field of fibre-optic temperature sensing, directional drilling techniques, three-dimensional freeze-pipe survey methods and the development of numerical calculation methods over recent years. Despite this, many ground freeze designs are still conducted through the application of analytic, semi-empirical design methods. Thus, opportunities for design optimization are often missed.

The development of the Ust-Jaiwa potash mine, located in the Perm Region of the Russian Federation, commenced in early 2012. It involves the construction of two vertical shafts with a final diameter of 8.0 m and a depth of approximately 500 m each. Both shafts are constructed through the application of the freeze shaft method with a freeze depth of 250 m. A freeze plant with a freeze capacity of 3.0 MW was installed to cope with local ground freeze challenges, including the freezing of salt-marl strata at a melting temperature of $-21°C$.

In this paper, conventional and advanced methods of freeze circle design are discussed in the context of their application for the Ust-Jaiwa freeze shaft project. The results of predictive calculations for the determination of frost development are outlined, and respective conclusions are drawn.

Moreover, the advantages of implementing numerical calculation methods for the back-analysis of thermal rock material parameters are demonstrated. The results of back-analysis are subsequently used for the predictive computations of frost development. It is concluded that state-of-the-art techniques for on-site monitoring and numerical analysis provide a significant potential for both the design and operation of large-scale freeze plants.

KEYWORDS: Freeze circle design, finite-element calculation, freeze shaft

1. INTRODUCTION

The freeze shaft method has been used in the mining industry since the late 1800s. During the period of shaft construction, this method provides temporary ground support and water control through artificial ground freezing.

Ground freezing is based on a simple physical concept. Brine with a specified temperature (below the freeze point of the water to be frozen) is pumped through a PE pipe inside of the freeze pipe. At the bottom of the freeze pipe, the brine escapes from the PE pipe and subsequently rises up through the annular space to the freeze head. The brine extracts energy from the rock mass around the freeze pipe. As a result, the temperature of the rock decreases and the temperature of the brine increase. The heated brine flows to the freeze plant and is chilled down to the desired temperature. Figure 1 shows a technical sketch of a freeze pipe and the brine circuit.

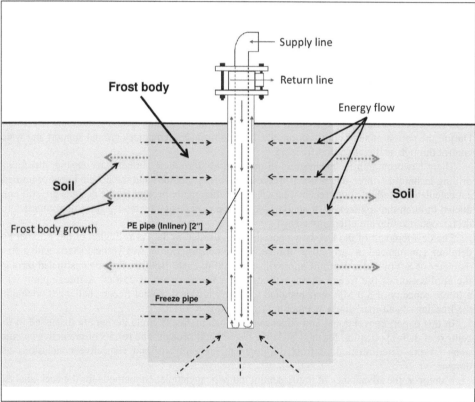

Figure 1. Schematic illustration of a closed brine circuit

The process of ground freezing can be split into two phases. The first phase (Phase 1) is the period of time that is required for closure of the frost body between two neighbouring freeze pipes. The second phase (Phase 2) is defined as the required period of time to achieve a defined wall thickness.

Freeze shafts are commonly predesigned through employment of so-called conventional design methods. The methods are limited to basic geometries such as circles and lines. Generally, conventional methods are not capable of considering special geometries such as irregular freeze circle shapes.

The focus of this paper is to compare the analytical and numerical methods for freeze circle design. Limitations and required boundary conditions for each method will be demonstrated. The comparison will be conducted in reference to a large-scale freeze shaft project, the Ust-Jaiwa project, located in the Russian Federation. For this project, both the analytical and the numerical method are implemented. Furthermore, the numerical modelling is used to back-analyse the thermal rock properties, to conduct a calibration of the model.

In this section, the design methods applied in the course of this paper are introduced. The analytical methods are the methods of Sanger & Sayles [1] and of W. Ständer [2], respectively. The numerical computing method chosen for the preparation of this paper is the finite-element program *Temp/W* of Geo-Slope International, Calgary, Canada [3]. The respective advantages and disadvantages are outlined.

1.1. *Sanger & Sayles*

The method of Sanger & Sayles is based on the design procedure of Neumann and Kahkimov [1], which is based on the following assumption:
- *"Isotherms move so slowly they resemble those for steady state conditions.*
- *The radius of the unfrozen (rock) affected by the temperature of the freeze-pipe can be expressed as a multiple of the frozen (rock) radius prevailing at the same time.*
- *The total latent and sensible heat can be expressed as a specific energy which when multiplied by the frozen volume gives the same total as the two elements computed separately." [Ständer 1976].*

Based on a stationary heat flow, the required period of closure of the frost wall between two neighbouring freeze pipes can be calculated. The required period of time is actually calculated without regard to the influence of neighbouring freeze pipes. Figure 2 shows a radial temperature profile around a freeze pipe during Phase 1.

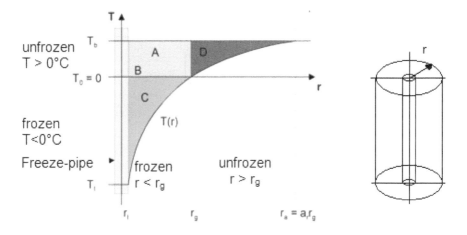

Figure 2. Radial temperature of the Sanger & Sayles model [1]

The temperature model shown in Figure 2 may be separated into two parts. The first part represents the frozen area around the freeze pipe. The second part describes the unfrozen area, which is already affected by the freeze pipe. This area is defined by a factor a_r. The results of the experiments conducted by Sanger & Salyes a_r equals "3" during Phase 1 [1].

The required period of time for Phase 1 can be obtained from Equation (1):

$$t_I = \frac{R^2 L_1}{4 K_1 v_s} \left[2 \ln\left(\frac{R}{r_0}\right) - 1 + \frac{C_1 v_s}{L_1} \right] \tag{1}$$

$$\text{with: } L_1 = L + \frac{a_r^2 - 1}{2 \ln a_r} C_2 v_0 \tag{2}$$

with:

t_I – Period of time for Phase 1 [days],
K_1 – Thermal heat conductivity of the frozen rock [W/mK],
C_1 – Heat capacity of the frozen rock [MJ/mK],
C_2 – Heat capacity of the unfrozen rock [MJ/mK],
L_1 – Volumetric latent heat of the unfrozen area [MJ/m³] (Phase 1),
R – Half the distance of the freeze pipe [m],
r_0 – Radius of freeze pipe [m],
v_s – Difference in the temperature of freeze pipe and freeze point of ground water [K],
v_0 – Difference in the temperature of the rock and freeze point of ground water [K],
a_r – Factor consider the sphere of influence.

Subsequent to Phase 1, the period of time for Phase 2 is calculated based on the assumption that an annular freeze design is chosen, both for the inner and outer area. The approach is based on the assumption that no heat energy is transferred into the area inside the freeze circle after the frost wall closure. Consequently, the frozen body develops quicker inside the freeze circle than on the outside. The equations for calculating the period of time t_{IIe} and t_{IIi} in dependence on the radius of the frost wall a and b are given below:

$$t_{IIe} = \frac{1}{2K_1 v_s} L_{II,e} \left[b^2 \ln\left(\frac{b}{R_p + \delta}\right) - \frac{b^2 - (R_p + \delta)^2}{2} \right] + \frac{C_1}{2K_1} \left[\frac{b^2 - (R_p + \delta)^2}{2}\right] \tag{3}$$

$$t_{IIi} = \frac{1}{2K_1 v_s} L_{II,i} \left[(R_p - \delta)^2 \ln\left(\frac{R_p - \delta}{a}\right) - \frac{(R_p - \delta)^2 - a^2}{2} \right]$$
$$+ \frac{C_1}{2K_1} \left[\frac{(R_p - \delta)^2 - a^2}{2}\right] \tag{4}$$

with:

$t_{II,e}$ – Period of time to close the frost body (Phase 2),
$t_{II,I}$ – Period of time to close the frost body (Phase 2),
A – Radius of the inner frost wall from the freeze pipe [m],
B – Radius of the outer frost wall from the freeze pipe [m],
R_p – Radius of the freezing circle [m],
δ – Radius of the freeze wall at the end of Phase 1 [m],
$L_{III,e}$ – Volumetric latent heat for the inner and outer area in Phase 2.

Because of the nature of these equations, an iteration process is required to achieve the optimum freeze circle design.

1.2. W. Ständer

The analytical method of W. Ständer is based on filamentary heat sink [2], in which the heat flow is considered to be constant over time. The calculation of Phase 1 is based on frost development around a single freeze pipe, analogous to Sanger & Salyes.

The influence of neighbouring freeze pipes is considered by the factor f_g. This factor reduces the initial temperature of the rock.

W. Ständer defines the dimensionless parameters in order to input parameters for nomographs, which are used to determine the required period of time for Phase 1.

For these nomographs, the following input parameters need to be determined:

$$X = -\frac{\lambda_1 t_I}{\lambda_2 t_{II}} \tag{5}$$

$$Y = -\frac{\lambda_1 t_I}{a_q \varrho q_s} \tag{6}$$

$$Z = \frac{R}{r_0} \tag{7}$$

$$k^2 = \frac{R^2 - r^2{}_0}{4a_1 \tau_g} \tag{8}$$

$$\beta = \frac{a_1}{a_2} \tag{9}$$

with:

λ_1, λ_2 – Thermal heat conductivity of frozen/unfrozen ground [W/mK],
t_I – Temperature of freeze-pipe [K],
t_{II} – Natural ground temperature [K],
r_0 – Radius of freeze pipe [m],
R – Radius of frost body from the freeze pipe [m],
a_1, a_2 – Thermal diffusivity of frozen/unfrozen ground [m²/s],
τ_g – Period of time for Phase 1 [s],
ϱ – Density of the soil [kg/m³],
q_s – Crystallisation of the formation water [kJ/kg K].

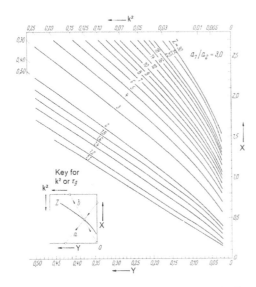

Figure 3. Nomograph for determining the frost development around a single freeze pipe [2]

The nomograph in Figure 3 provides the parameters W, depending on the ratio of a_1 and a_2. With the parameter W it is possible to calculate the period of time in Phase 1, compare Equation (10).

$$\tau_s = \frac{R^2 - r^2}{4a_1 * k} * m_s \tag{10}$$

with:

τ_s – Period of time for Phase 1,
m_s – Correction factor of the closing time = 0.3.

Subsequently, the average temperature on the freeze circle is calculated, and the wave-like freeze wall is flattened. The area to the outer and inner direction of the freeze circle is underestimated, and the area in the direction of the neighbouring freeze pipes is overestimated.

The actual determination of Phase 2 is made by the multiplication of Phase 1, which considers only the heat of crystallisation τ_0 and the ratio, and includes the pre-cooling in plane and the influence of the cylindrical shape. The necessary proportional factors to calculate the effect of the unfrozen area of the pre-cooling heat can be determined by the nomographs, which are shown in Figures 4, 5 and 6.

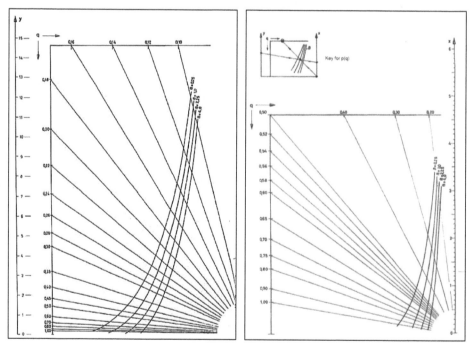

Figure 4. Nomograph for determining the factor q [2]

Figure 5. Nomograph for determining the factor q [2]

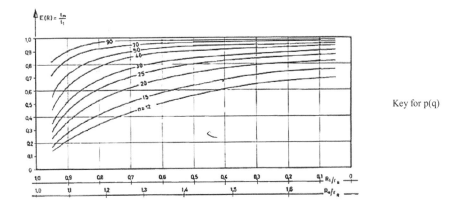

Figure 6. Nomograph for determining of the ratio E(r) [2]

The factors p_v and p_0 can be calculated by the Equations (11) and (12).

$$p_v = \sqrt{4 * a_1} * q \; ; \; t_I = t_m \text{ and } t_{II} = natrual \text{ } ground \text{ } temperature \tag{11}$$
$$p_0 = \sqrt{4 * a_1} * q \; ; \; t_I = t_m \text{ and } t_{II} = 0°C \tag{12}$$
$$t_m = E(r) * t_I \tag{13}$$

with:

> p_v –Proportional factor for pre-cooling heat,
> p_0 –Proportional factor including exclusive consideration of crystallization,
> t_m –Average freeze pipe temperature,
> $E(r)$ –Ratio of the average temperature on the freeze pipe circle to the freezing
> temperature depending on the frost body radius.

As a result of the different geometries and effect of heat on the inner and outer area of the freeze circle, the calculation will be separated for the two areas [Equations (14) and (15)].

$$\tau_{v\,a} = \frac{p^2_0}{p^2_v} * \left[1 + \frac{\frac{\lambda_2 t_{II}}{2R_m \varrho q_s}(R_a\text{-}r_R)}{p^2_v} + \frac{2(\frac{\lambda_2 t_{II}}{2R_m \varrho q_s})^2 (R_a\text{-}r_R)^2}{p^4_v} \right]^2 * \tau_0 \tag{14}$$

$$\tau_{v\,i} = \frac{p^2_0}{p^2_v} * \left[1 - \frac{\frac{\lambda_2 t_{II}}{2R_m \varrho q_s}(r_R\text{-}R_i)}{p^2_v} + \frac{2(\frac{\lambda_2 t_{II}}{2R_m \varrho q_s})^2 (r_R\text{-}R_i)^2}{p^4_v} \right]^2 * \tau_0 \tag{15}$$

with:

> $\tau_{v,a}$ –Period of time to close the frost body [days],
> $\tau_{v,i}$ –Period of time to close the frost body [days],
> R_m –Temporary average of the frost body radius [°C],
> τ_0 –Period of time to close the frost including exclusive consideration
> of the crystallization [days].

95

1.3. *Numerical modelling*

Because of the computing power available today, complex geotechnical problems can be solved through numerical modelling. For this paper, the two-dimensional finite-element program *Temp/W* is used. This software is designed for solving thermal problems [3], and it is based on the following two-dimensional heat balance [Equation (16)]:

$$\frac{\partial}{\partial x}\left(k_x\frac{\partial T}{\partial x}\right)+\frac{\partial}{\partial y}\left(k_y\frac{\partial T}{\partial y}\right)+Q = \lambda\frac{\partial T}{\partial t} \tag{16}$$

with:

T – Temperature,
k_x – Thermal conductivity in x-direction,
k_y – Thermal conductivity in y-direction,
Q – Initial heat flow,
λ – Thermal capacity,
t – Time.

Equation (16) describes how the heat inflow and outflow for a sufficiently small time step of an element is equal to the change in heat within an element.

The *Temp/W* solver employs commonly used numerical methods to partition the problem, resulting in a set of equations that can be used when implementing mathematical procedures that are not discussed in this paper.

1.4. *Comparison of the analytical and numerical design*

Table 1 gives a comparison of the analytical and numerical methods introduced previously.

Table 1. Comparison of the analytical and numerical methods for investigation of frost development

	Analytical method		Numerical method
Methods	Sanger & Sayles	W. Ständer	*TempW*
level of difficulty	+	+	−
amount of work	+	+	−
time exposure	0	0	−
requirement	+	++	0
customisability	−	−	++
necessary input	+	+	+
depth of the process	0	0	+

++: Very good/very simple/very low;
+: Good/simple/low;
0: Neutral;
−: Bad/difficult/high;
—: Very bad/very difficult/very high.

2. GROUND FREEZING AT THE UST-JAIWA PROJECT

The development of the Ust-Jaiwa potash mine, located in the Perm Region of the Russian Federation (Fig. 7) commenced in early 2012. It involves the construction of two vertical shafts with a final diameter of 8.0 m and a depth of approximately 500 m each. Both shafts are constructed through the application of the freeze shaft method with a freeze depth of 250 m.

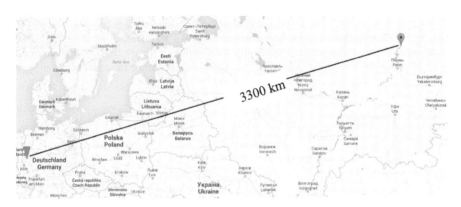

Figure 7. The Ust-Jaiwa project, located in the Russian Federation [Google Earth]

The region is located in the east European plain with an average height of 100 to 300 m asl. The geological strata are plate-shaped. There are five fundamental strata from ground level to a depth of 465m. These formations are as follows:
- Quaternary formation (loose sediments);
- Coloured formation (various colour clay, argillite, and sandstone);
- Terrigenous-carbonat formation (primarily fractured hard rock of dolomite, limestone, and marl);
- Salt-marl formation (primarily marl, clay, and limestone); and
- Salt formation (carnallite and kama salt).

Unconfined water is located within the coloured formation. Strained water is also located in the terrigenous-carbonat formation. No water is found in the lower formations.

The deepest freezing point was determined by a laboratory test at the salt-marl formation, with a freeze point of −21°C.

In this case, a freeze plant with a freeze capacity of 3.0 MW was installed. The freeze circle was installed with a diameter of 17.0 m. Forty-five freeze pipes were installed for each shaft with a spacing of 1.19 m and a depth of 245.0 m. Furthermore, four measurement holes were drilled with a depth of 245.0 m.

2.1. Drilling method and drilling measurement

The holes were drilled using two mobile drill jumbos. Two different systems were used during the drilling to reach the required accuracy in verticality. One system is called "Measure while drilling" (MWD), which allows constant measuring of the drilling operations. The second system is called the "Motary steerable directional control system", which allows full control of the direction of the drill head.

In addition to the MWD, every hole was measured with a gyroscope after the drilling operation. As a result of these measurements, it the exact course of every hole was able to be determined.

2.2. Temperature measurement method

Also critical for a successful freeze shaft project is the temperature measurement. By measuring the rock's temperature, a conclusion can be made about the freeze growth. To obtain a meaningful result, it is necessary to use a high-quality measurement system, such as fibre-optic cables. A well-known thermal characteristic of fibre-optic cable is its ability to measure the temperature by sending a laser pulse through the cable. The system records the backscatter intensity and the time taken for the laser pulse. With the backscatter intensity it is possible to measure the temperature over a defined space (in this case, 0.5 m). The determination of the temperature is based on the Stokes/ /Anti-stokes line. The temperature resolution amounts to +/−0.3 K. Because the fibre-optic cable measures the entire depth, it is necessary to identify the position of the measured temperature. By recording the time that the laser pulse is needed, it is possible to determine the exact position.

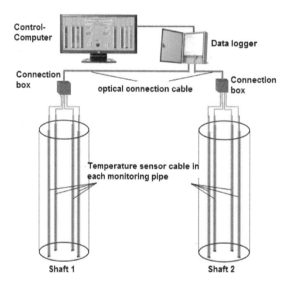

Figure 8. Schematic construction of the measurement system

This implies that in the shaft freeze project, a closed grid could be built. These data were collected at a splice box and sent to a data logger, which is linked to a computer and connected to the Internet. In this way it is able to control the data from anywhere, compare Figure 8.

3. APPLICATION OF THE METHODS AT THE FREEZE SHAFT PROJECT UST-JAIWA

The chosen methods were used because of the boundary condition, with particular focus on the applicability and optimizing of the freeze shaft process before and during the process.

In the context of this paper, the methods are employed at the salt-marl strata. The used input parameters are provided in Table 2. In the following chapter the methods will be calculated, followed by the results and validity checks.

Table 2. Summary of Input Parameters

Technical parameters	Input	Unit
Radius of freeze pipe	0.06985	m
Half distance between two freeze pipes	0.595	m
Planned strength of the freeze wall	3.3	m
Radius of freeze circle	8.5	m
Number of freeze pipes	45	Rohre
Temperature at freeze pipe wall	-35	°C
Parameter of the rock		
Initial rock temperature	6.0	°C
Thermal conductivity of the rock (frozen/unfrozen)	8.80/5.72	kJ/hm K
Heat capacity of the rock (frozen/unfrozen)	2.231/2.878	MJ/m³ K

3.1. Sanger & Sayles

As stated previously, the following equations are obtained:

$$t_I = \left[\frac{0.354025 * 343.3}{0.04076 * 14}\right] * \left(4.2844 - 1 + \left[\frac{2.264 * 14}{343.3}\right]\right) = 719 \ [h] \tag{17}$$

$$t_{II,i} = \frac{1}{2*8.80*14} *275.76* \left[(8.5\text{-}0.47)^2 * \ln\left(\frac{8.5\text{-}0.47}{5,2}\right) - \frac{(8.5\text{-}0.47)^2\text{-}5.2^2}{2}\right] + \frac{2.231}{2*8.80}$$
$$* \left[\frac{(8.5\text{-}0.47)^2\text{-}5.2^2}{2}\right] = 12{,}778 \ [h \tag{18}$$

$$t_{II,e} = \frac{1}{2*8.80*14} *301.28* \left[11.8^2 * \ln\left(\frac{11.80}{8.5+0.47}\right) - \frac{11.80^2\text{-}(8.5+0.47)^2}{2}\right] + \frac{2.231}{2*8.80}$$
$$* \left[\frac{11.80^2\text{-}(8.5+0.47)^2}{2}\right] = 14{,}475[h] \tag{19}$$

These equations show that it is quite simple to calculate the time for Phases 1 and 2. In this case the results represent only the time for one defined freeze wall radius. If it is necessary to get results for more than this radius, the calculation must be repeated. It should be noted that the factor a_r is determined solely by experiment with soil. Sanger & Sayles set this factor to 3 at Phase 1 and between 4 and 5 at Phase 2.

3.2. W. Ständer

The method of W. Ständer is used by nomographs, which simplifies the calculations by hand, but complicates the use of calculation programs like *Microsoft Excel.* In the context of these calculations, the authors' own program was written to solve the methods using a calculation program. In doing so, the program was able to solve many variants, including the freeze process by iteration.

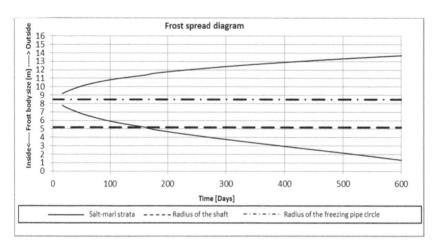

Figure 9. Representation of frost development over time according to the method of W. Ständer

Figure 9 shows the results for the salt-marl strata. The continuous lines represent the freeze wall. Between the two lines the temperature is below the freezing point of water. After 176 days, the line crosses the shaft diameter of 5.2 m. This point describes the planned freeze-wall thickness.

3.3. *TempW*

The application of the two-dimensional finite-element program requires thorough and careful planning and preparation. In addition to the thermal parameters, the exact position of all pipes and freezing temperature measuring holes were determined and entered into the simulation. Moreover, the exact continuous measurements of brine temperatures and time were entered.

In addition to these input parameters that are necessary for the calculation, a timing diagram was developed to represent all construction events. It provides a construction site event as a state change of the boundary conditions. Table 3 shows this preparation as an example.

Table 3. Schedule of Site Events

No.	Shaft 1		
	Cause	Execution	Time period
1	Start of freezing measure	18.08.2013	Instantly 18.08.2013
2	Thermal calculation		18.08.2013 – 03.12.2013
3	Failure (Freeze pipe): 28, 31, 51	03.12.2013	Instantly 03.12.2013
4	Thermal calculation		03.12.2013 – 23.12.2013
5	Integrate (Freeze pipe): 51	24.12.2013	Instantly 24.12.2013
6	Thermal calculation		24.12.2013 – 27.12.2013
7	Integrate (Freeze pipe): 28	28.12.2013	Instantly 28.12.2013
8	Thermal calculation		28.12.2013
9	Integrate (Freeze pipe): 31	29.12.2013	Instantly 29.12.2013
10	Thermal calculation		29.12.2013 – 20.01.2014
	… etc.		
	End of the simulation		

In addition to determining the frost body development, the process is used to calibrate the entire system based on existing data. As a result, the calculated temperatures were iteratively compared with the present. When insufficient compliance of the temperature gradients is determined, the calculation must be repeated by adjusting the input parameters. For a given geometry and known initial parameters (e.g., the soil's initial temperature), it is possible to recalculate the rock mass parameters, and thus a model calibration can be achieved, compare Figure 10.

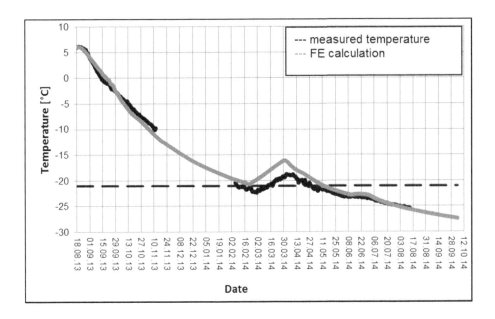

Figure 10. Measured and predicted temperature development over time after model calibration

Excerpts of the recalculated parameters compared with previous studies are given in Table 4.

Table 4. Comparison of Recalculated and Given Rock Parameters

Parameter	Recalculated parameter	Values from geotechnical report
Water content [%]	10.0	16.4
Thermal conductivity (frozen/unfrozen) [kJ/hm K]	10.04/6.83	8.8/5.72
Heat capacity (frozen/unfrozen) [MJ/m³ K]	2.231/2.878	2.231/2.878

The back-calculated soil parameters clearly show only small differences in consideration of the thermal properties. However, the information on the water content has significant differences, sug-

101

gesting that a general representation of soil parameters over an entire shift in special cases is not sufficient. Furthermore, it shows that a calibration, in this case, is necessary and targeted by a re-calculation of the rock parameters. By a precise calibration of the numerical system, a meaningful and stable result can be achieved.

Based on the previous calculations, a stable system can be created, which depicts the tempera-ture profile closely to reality. Consequently, the interpretation of the obtained results and display of the frost body at any time can be generated.

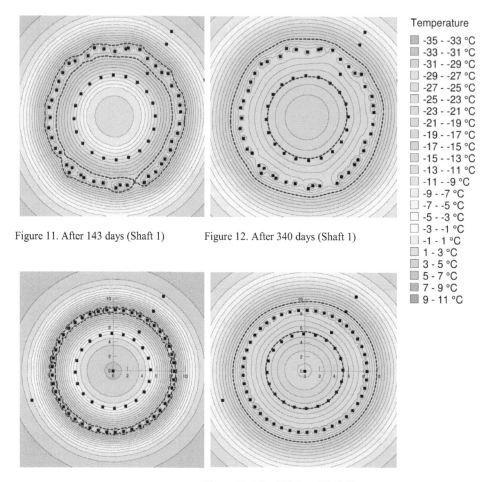

Figure 11. After 143 days (Shaft 1)　　　Figure 12. After 340 days (Shaft 1)

Temperature

■ -35 - -33 °C
■ -33 - -31 °C
■ -31 - -29 °C
■ -29 - -27 °C
■ -27 - -25 °C
■ -25 - -23 °C
■ -23 - -21 °C
■ -21 - -19 °C
■ -19 - -17 °C
■ -17 - -15 °C
■ -15 - -13 °C
■ -13 - -11 °C
■ -11 - -9 °C
■ -9 - -7 °C
■ -7 - -5 °C
■ -5 - -3 °C
■ -3 - -1 °C
■ -1 - 1 °C
■ 1 - 3 °C
■ 3 - 5 °C
■ 5 - 7 °C
■ 7 - 9 °C
■ 9 - 11 °C

Figure 13. After 80 days (Shaft 2)　　　Figure 14. After 300 days (Shaft 2)

The upper illustrations, Figures11–14, show the temperature contours for Shaft 1 and Shaft 2 for different points in time in a depth of –230 m. The blue dashed line represents the frost line. The circular points represent the future shaft lining, and are used in these drawings only as a guide.

These results show the influence of the inclusion of current daily updated measured values and the exact freeze pipe location at a defined depth. The freezing pipe arrangement for the time of closing is clearly a decisive factor.

4. DISCUSSION OF RESULTS

The methods used result in significant differences in the closing and pre-freezing times. Thus, the selection of a method is a crucial factor in the pre-planning and dimensioning of a freeze shaft project, as it serves as a basis for calculation.

When considering the closing time, an underestimation of frost development can be identified irrespective of the method used. In both cases, for Shafts 1 and 2, the closing time is much longer than predicted. When looking at the pre-freezing time, the method of Sanger & Sayles overestimated the process and the method of W. Ständer and *TempW* underestimated the process. The results of the calculations are provided in Table 5.

Table 5: Summary of the Results

		Salt-marl strata	
		Closing time	Pre-freezing time
Ideal	W. Ständer	8.1	176.9**
	Sanger & Sayles	39.5*	532.0
	Temp/W	20	220
(Shaft 2)	*Temp/W*	80	300
(Shaft 1)	*Temp/W*	143	340
*Effects resulting from adjacent freeze pipes were not taken into account.			
**Until a total wall thickness of 5.79 m.			

This underlines the influence of the boundary conditions and demonstrates the absolute necessity of flexible methods to calculate the freeze process.

The results for the pre-freezing time show that the different approaches and methods of calculation have a significant effect on the determination of time. Compared with the numerical calculation, the method of W. Ständer is regarded as optimistic; in contrast, the Sanger & Sayles method offers a more conservative estimate.

By accepting the results that were obtained using the present project data, the closing times' estimations are too low in all cases. This error results from the assumption of ideal state conditions, which are usually not present. Consequently, an interpretation of the ideal results is essential and can only be used as reference values. The freezing time shows that the conservative approach of the Sanger & Sayles method is on the safe side.

CONCLUSIONS

In the paper, the differences amongst the various calculation methods are well demonstrated, and their advantages and disadvantages are identified. It is shown that the estimation of the closing and pre-freezing times with many uncertainties is possible as a result of idealised assumptions. All methods overestimated the closing time. A safe estimate is therefore only possible if all constraints and geometries of a project are known, and the process can deviate from the idealised approach. In addition, it was found that determining the time required until the configured frost wall thickness is reached depends on the chosen approach.

Thus, it can be concluded that the results achieved using analytical methods and when assuming idealised boundary conditions can only be used as an indication. More accurate investigations of frost development are possible with the use of modern finite-element model simulations.

REFERENCES

[1] Sanger F.H. & Sayles F.J. 1979: Thermal and Rheological Computations for Artificially Frozen Ground Construction. Engineering Geology, Vol. 13, Issues 1–4, Pages 95–117.
[2] Ständer W. 1967: Mathematischer Ansätze zur Berechnung der Frostausbreitung in ruhendem Grundwasser im Vergleich zu Modelluntersuchungen für verschiedene Gefrierrohranordnungen im Schacht- und Grundbau. Institute für Bodenmechanik und Felsmechanik Technische Hochschule Karlsruhe.
[3] Geo-Slope International 2012: Geo-Studio 2012 Software Package. Calgary, Canada.

International Mining Forum 2015, Kicki et. al. (eds) © 2015 Taylor & Francis Group, London, UK. ISBN 978-1-138-02820-3

Static Calculations of Mine Shaft Linings
in Poland (Selected Problems)

A. Wichur, K. Frydrych, P. Kamiński
AGH – University of Mining and Metallurgy, Krakow, Poland

ABSTRACT: Long-lasting prosperity period of mining industry after the Second World War resulted in intensive development of the underground mine building. This building was exposed to numerous tasks like building and extension of mines in worse and worse geological conditions, including increasing thickness of the water saturated overburden and increasing exploitation depth. Shaft sinking in more and more difficult hydro-geological conditions (big thickness of the water saturated overburden) resulted in necessity of elaboration and implementation of new constructions of water-proof lining of high load-bearing capacity. It was also proved that using classic formulas of shaft lining calculation resulted in too excessive thickness of the lining in question. Thus realization of suitable investigations was necessary. Technical and technological experience gained in tests and mining operations allowed improvement of the design methods and development of suitable technologies. It particularly concerned methods of the shaft lining design. This study was aimed at the acquainting the problem in question to specialists – designers that design of the underground headings (particularly shafts) comprises a process of development of rules and procedures accompanying broadening knowledge of the rock mass and lining material properties. The examinations are summarized with conclusions indicating directions of further development of the shaft lining design.

KEYWORDS: Shaft lining, lining design

1. INTRODUCTION

Establishment of mining companies specialized in mining building was a consequence of the implementation of mineral resources and energy based programs, considered as strategic industrial branch for a long time named as national industry [Wichur, Żyliński 2000]. In the period 1952––1994, 24 hard coal mines, copper ore mines and salt mines, have been built in Poland 9, and completely new mining regions such as Rybnik Coal Mining Region, Legnica-Głogów Copper Region or Lublin Coal Basin, have been created. In order to realize the mentioned investment tasks a special research-design background has been developed, as well as new disciplines were introduced into mining universities to teach students problems related with mine design and mine building. These activities allowed implementation of broad range of investment works, often in very hard geological – mining conditions, for example in the period 1945–1989, and in total 260 km of shafts and pit-holes have been sunk. In favorable market conditions for mining works, in the sixties and seventies mining work companies employed in total about 40 thousands workers, what amounted for 8 to 10% of all workers employed in underground mining [Wichur, Żyliński 2000].

The gained technical and technological experience allowed development of design methods and improvement of the technology. In particular it was related with shaft lining design methods. The present study is aimed at explanation these problems to designers, including demonstration that design of underground excavations lining (particularly shaft lining) should be considered as a process of modifications of procedures accompanying gathering knowledge about properties of the rock mass, including lining material. The discussion is completed with suitable conclusions, which determine directives of further directions of the development of shaft lining design methods.

2. TRADITIONAL METHODS OF THE SHAFT LINING CALCULATIONS

As shafts sinking are executed in more and more difficult hydro-geological conditions (high thicknesses of water saturated overburden) it was a need of development and implementation of new water-proof lining structures having high load-bearing capacity [Olszewski, Kostrz 1995]; [Tajduś, Wichur 2009]. It was also proved that in such conditions using standard formulas for calculation of the shaft lining thickness [Kostrz 1964]; [Borecki et al. 1959]; [Ustawa 1993] resulted in excessive thickness of the designed lining. Cymbariewicz's and Huber's formulas were commonly used in practice (for lining thickness calculation) [Wichur 1972b]. Commonly used Cymbarewicz's formula was derived on the basis of the analysis of so called "retaining wall scheme", which is known in soil mechanics and rock mass mechanics (Fig. 1). Among formulas used we can mention for example (formulas for uniform rock mass):

$$p = \gamma \cdot H \cdot tan^2 \left(45^\circ - \frac{\varphi}{2} \right) \tag{1}$$

and:

$$p = \gamma \cdot H \cdot tan^2 \left(45^\circ - \frac{\varphi}{2} \right) - 2,5 f \tag{2}$$

where: p = retaining wall loading (shaft lining loading), MPa; γ = volume weight of the rock mass (soil), MN/m^3; H = distance between retaining wall cross-section (for which the load is calculated) and ground surface, m; φ = apparent internal friction angle of the rock, deg; f = coefficient of internal friction of the rock.

$$\varphi = arc \, tan \, f = arc \, tan \frac{R_{cs}}{10} \tag{3}$$

where R_{cs} = rock compressive strength.

Formula (2) additionally comprised influence of the rock compressive strength onto load reduction.

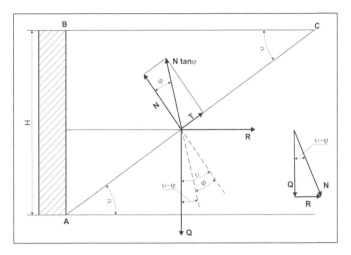

Figure 1. Retaining wall scheme

Commonly used Huber's formula (for calculation of the lining thickness) is a kind of implementation of the theory of elasticity used for solution of this problem. Use of traditional formulas was not successful, thus suitable research works were necessary. Such works have been executed in a number of research centers, and the most known results were achieved in the Department of Research and Experimental Studies of Mine Building (Zakład Badań i Doświadczeń Budownictwa Górniczego) [Wichur 1971c]; [Wichur 1971d]; [Wichur 1972a]; [Wichur 1972d]. These works have been implemented in the drafts of branch standards [BN-71/0434-02]; [BN-72/0434-03]; [Chudek et al. 2013] – the drafts have been elaborated by the team: PhD, Eng. Jan Kostrz; PhD, Eng. Ryszard Majchrowicz; PhD, Eng. Eugeniusz Posyłek and M.Eng. Andrzej Wichur. In time, this method was exposed to practical verification and minor changes [Wichur 1986]; [Wichur 1996], taken into consideration suitable standards (BN-83/0434-02; BN-79/0434-03). After resolution about the standardization act [Ustawa 1993] the verification of branch standards has been executed and two new Polish standards were implemented [PN-G-05015:1997], [PN-G-05016:1997] – prof. Andrzej Wichur was the author of drafts of these standards, and continuity of the design and using of the shaft lining design rules were kept. Long-term period of using the standards in questions has proved their usability for elaborated calculation method.

3. STATIC CALCULATIONS OF SHAFT LINING ACCORDING TO POLISH STANDARDS

3.1. *Calculation of the lining loadings*

Formulas (1) and (2), which have been commonly used since decades resulted in some habits of the designers being the result of design routine, thus it was decided that they should only be tailored to actual knowledge level, as well as to design and sinking of the mine shafts. Implementation of the study proposed by the author [Wichur 1970a] describing the concept of the characteristic (standard) and design loading of the shaft lining – have been considered as the major assumption.

Characteristic (standard) loading of the shaft lining (at the depth H) has been determined [BN-71/0434-02]; [PN-G-05016:1997]; [Wichur 1970a]; [Wichur 1971c]; [Wichur 1971d]; [Wichur 1972d] as a uniform load of the horizontal shaft section equivalent (in sense of the effort of this shaft section) to action of the mean value of the loading in average conditions. This definition allowed combining the elaborated standards [BN-71/0434-02]; [BN-72/0434-03]; [PN-G-05015: :1997]; [PN-G-05016:1997] with a system of Polish standards [PN-64/B-03001]; [PN-76/B-03001] and also allowed adaptation of traditional formulas for calculation of the lining dimensions [Wichur 1972b]. This in turn allows application of measurement results "in situ" for purposes of design loads of shaft lining, what will have essential meaning in the future; suitable procedural steps in such cases have been discussed in details in the paper [Wichur 1972d].

After introduction of the concept of the characteristic load, it was easy, according to standards [PN-64/B-03001]; [PN-76/B-03001], to use load (overload) factor as a coefficient giving consideration to probability of the occurrence of loads, which are less favorable than the characteristic loads. Similar procedure was followed in case of the introduction of the design load as the product of the characteristic load and load (overload) coefficient [Wichur 1972b].

Introduction of the concept of the critical depth [Wichur 1972b] (Fig. 2) allowed avoidance of the excessive lining dimensions in the rocks at small depths – diversification of the load coefficient values – allowed taking into account the influence of the shaft diameter, distance of the shaft cross section from the shaft station entrance, as well as un-stressing action of strongly cohesive layers onto located between them weakly cohesive layer, and their influence of shaft lining load, has been proved during measurements [Wichur 1971c]; [Wichur 1971d]; [Wichur 1972d]. It should be noted that proposed concept of critical depth was then used in design of the shell and vaulted linings [BN-79/0434-04; BN-78/0434-07]. Phenomenon of the lining load stabilization at big depths (so called boundary depth, important particularly in coal seams occurring at high depths), was also taken under [BN-83/0434-02]; [PN-G-05016:1997]; [Wichur 1986].

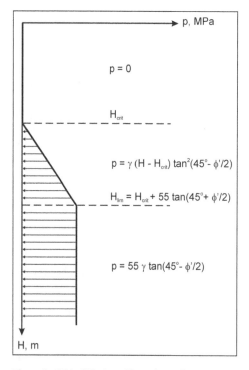

Figure 2. Critical Hcrit and boundary Hlim depth
in the rock mass around the shaft

108

Actually, particularly important is the procedure of calculation of the shaft lining load below the boundary depth. This procedure can be used if three conditions are satisfied:

1. shaft section is located below the critical depth,
2. the part of the rock mass profile, in which given shaft segment is designed, is classified into I or II degree of the water hazard (i.e. there is no possibility of the water entrance into shaft with sine-grained rock material),
3. the ratio compressive strength/tensile strength satisfies the condition.

$$\frac{R_{cs}^{(n)}}{R_{rs}^{(n)}} > \begin{cases} \left[\dfrac{1-v}{v\,(1+v\,)}\right]^2 \\ \left[\dfrac{2}{1+v}\right]^2 \end{cases}$$

(4)

adequately for $v \le \dfrac{1}{3}$ and $v \ge \dfrac{1}{3}$,

then the characteristic (standard) load can be calculated from the formula:

$$p_N = 55\,\gamma_{sr}^{(n)}\,tan\left(45^\circ - \frac{\Phi^{(n)}}{2}\right)$$

(5)

where: $R_{rs}^{(n)}$ = characteristic value of the rock tensile strength, MPa; $R_{cs}^{(n)}$ = characteristic value of the rock compressive strength, MPa; v = rock Poisson's coefficient; $\Phi'^{(n)}$ = characteristic value of rock effective internal friction angle, deg; $\gamma_{sr}^{(n)}$ = characteristic value of the mean volume weight of the overlying rocks, MN/m^3.

In water saturated rocks one more condition must be satisfied – resistance for the water action (A according to Skutta's scale or $r = 1$ according to GIG [BN-79/0434-04].

Formula (5) is a semi-empirical formula, derived on the basis of the boundary depth assumption and results of the observation of shaft lining condition in hard coal mines [Wichur 1996]. In comparison with the previous standard [BN-83/0434-02], following modification from the year 1990, the value of the numerical coefficient in formula (5) has been reduced from 70 into 55, considering observation of fast weakly cohesive rocks un-stressing according to advance of the shaft face. Inequality (4) corresponds to condition of the appearance of parting fractures in the rock mass around the shaft [Wichur 1986].

A rule [BN-83/0434-02]; [PN-G-05016:1997] that value of this pressure is calculated on the basis of supposed hydro-geological conditions after finishing the shaft sinking, taking under consideration shaft sinking technology (rock mass freezing, preliminary plugging or rock mass drainage, as well as use of the concrete lining in layers with poor permeability) was introduced in calculation of the water pressure.

3.2. Calculation of the lining thickness

Huber's formula modified by the main author of the presented paper [Wichur 1971c]; [Wichur 1971d]; [Wichur 1972d] has been proposed for calculation of the single concrete lining:

$$d_b = a \cdot \left(\sqrt{\frac{R_{bb}}{R_{bb} - m \cdot p \cdot \sqrt{3}}} - 1 \right) \tag{6}$$

where: d_b = the thickness of the shaft single concrete lining, m; a = the internal diameter of the shaft, m; p = design loading of the shaft lining, MPa; R_{bb} = design compressive strength of plain concrete, MPa – acc. to standard [PN-84/B-03264] (at present: f_{cd}^* acc. [PN-B-03264:2002], MPa; m = corrective coefficient taking into consideration the load ununiformity (m = 0,95–1,15).

In case of shafts sunk with use of the rock mass freezing method the assumed thickness of the concrete shaft lining shouldn't be less than minimal value 0,35–0,50 m, depending on the temperature of the frozen wall.

Similar formula was applied for calculation of the lining built from bricks and concrete blocks, whereas in this case calculation of the lining thickness with use of the permissible stress was recommended. In the last version of the standard [PN-G-05015:1997] it was suggested that following the examinations [Wichur el al. 1991] the following formula for the design strength of the wall built of concrete segments (segmental shaft lining) should be used:

$$R_p = 0,5\, R_{bb} \tag{7}$$

where: R_p = the design compressive strength of the wall built of concrete segments, MPa; R_{bb} = design compressive strength of plain concrete , MPa – acc. to standard [PN-84/B-03264] (at present: f_{cd}^* acc. [PN-B-03264:2002]), MPa.

In case of the calculation of multilayer shaft lining, a commonly used in design practice loading distribution was assumed: load from the water side acts onto internal (water-proof) lining ring (for example concrete with insert of hydro-insulating foil), and load from side of rock mass acts onto external lining ring (preliminary lining). Analysis of the load bearing capacity of the shaft lining equipped with non-linearly elastic material has been conducted in the study [Czaja 2002].

In case of the combined shaft lining built of non-metallic materials (for example brick, concrete blocks, concrete, reinforced concrete) classic elasticity theory formula was applied, and in case of combined tubing-concrete and steel-concrete lining – modified by domestic designers formulas F. Mohr's [Mohr 1946]; [Walewski 1965] were used. Comprehensive explanation of the calculation procedure of reinforced concrete lining has been described in the study [Wichur 1972b]. Values of mechanical parameters of the lining are selected from suitable Polish standards used for building structures.

More detailed information concerning calculation procedures of the shaft linings can be found in the study [Wichur 1996]; whereas chapter 3.3 is devoted to description of the probabilistic calculation model.

3.3. Probabilistic models used in shaft lining designing

Assumption of suitable load configuration (for example uniform load distributed on the shaft circumference) and its proper value plays important role in mine excavations designing. In the last century measurements of the shaft load "in situ" were started. These measurements proved that the load of underground excavation is characterized with specific non-uniformity (see [Krupenikov et al. 1966] (Fig. 3), which cannot be defined with use of commonly applied mathematical functions (for example model of the non-uniform load proposed by F. Mohr [Mohr 1950]).

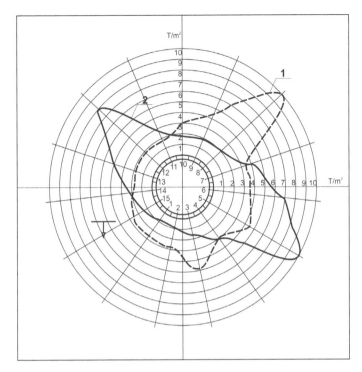

Figure 3. Example of results of measurements of a shaft lining loading
[Krupenikov et al. 1966] (2 – stated loading)

Linings of the underground excavations work in the rock mass, i.e. in medium in which the structure and parameters are par excellence variable and random, thus some trials have been undertaken to consider the un-homogeneity as a probabilistic model, that is a model taking under consideration random variability of the input variables in statistical calculations.

Practice proved usability of two such probabilistic models:

1. random variable model,
2. random function (stochastic process) model.

Theory and use of the probabilistic models in engineering structures design is presented in the study [Murzewski 1989]. Omitting unnecessary details of the structure of mentioned models (e.g., [Smirnow 1966]; [Szczepankiewicz 1985]; [Wichur 1970b]; [Wichur 1972a] we can state that some random variables are determined in so called probabilistic space, and are realized by numbers (for example results of geotechnical parameters denotations, results of building materials resistance denotations, etc.), whereas random functions of single independent variable (this limitation is reasonable in this case) can be considered as some sets of random variables, i.e. numeric-numeric functions of single independent variable in this case. Thus there are obvious possibilities of the successful application of these models – random variable model can be used for description of the material properties, which can be expressed in numerical manner, including description of the stable configuration load of the structure, whereas random function model can be used for example for the description of the lining load, which is variable in space (for example on the shaft circumference) and in time (loads obtained in measurements can be considered as realizations of the definite random function) (cf. Fig. 3).

111

The random function model has more complex structure and bigger possibilities of application. In Poland the model in question was applied during developing shaft lining calculation method [Wichur 1971a]. It required solution of the following problems:
– stress field and effort of the cross-section of the shaft loaded with stationary random load function [Wichur 1986]; [Wichur 1971b],
– probability of the destruction of shaft lining loaded with random function of the rock mass pressure [Wichur 1972c].

Analysis of the observed material obtained during measurement of shaft lining loads conducted with use of the presented model allowed formulation of the new formula used for calculation of the shaft lining thickness, which is based on probability of destruction of the shaft lining (Fig. 4) [Wichur 1971c]; [Wichur 1971d]; [Wichur 1972d], which was introduced to suitable branch standard [BN-72/0434-03] and is at present used [PN-G-05015:1997]; [PN-G-05016:1997].

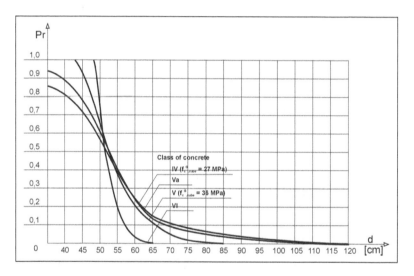

Figure 4. Dependence of the probability of the shaft concrete lining destruction on the lining thickness (numerical example) [Wichur 1971a]

The model in question has been described in the study [Wichur 1976]. Problem of the application of random function model in calculation of the tunnels lining, has been discussed in the study [Šejnin 1969].

The second probabilistic model (random variable model) was used in the study [Duży 2007] for the evaluation of the dog heading steel arch support and rock mass interaction. Parameters of the probability distribution of two random variables: support loading and its load bearing capacity are considered as function of random variables (geo-technical, mining and material parameters). The obtained results were used for the development of classification of dog headings stability conditions.

Random variable model was also used in the study [Nguyen, Wichur 1985] to define relation between destructive bending moment and longitudinal destructive force in case of eccentric compression, taking under consideration random strength of the concrete. In the study [Domańska 2003] the influence of the random variability of steel strength parameters onto load bearing capacity of the steel arch support made of V-shaped section was examined with use of the similar model.

SUMMARY AND CONCLUSIONS

Long-lasting prosperity period of mining industry after the Second World resulted in intensive development of the underground mine building. This building was exposed to numerous tasks like building and extension of mines in worse and worse geological conditions, including increasing thickness of the water saturated overburden and increasing exploitation depth. However this task could be realized under condition of the technology development, including development of the new mine lining technologies and design methods. Numerous new technologies of long-lasting linings of underground roadways, including design methods, have been developed and number of new calculation methods, including probabilistic models, has been implemented. Long-lasting use of these methods proved their usability not only in conditions of the underground mining but also in underground building not related with mining industry [Rułka et al. 1996]. Further development of the design methods related with long-lasting linings of underground mining roadways (including mine shafts) should go in the direction of broader application of geotechnical measurements "in situ".

REFERENCES

BN-71/0434-02 Szyby górnicze – Obudowa – Obciążenia.
BN-72/0434-03 Szyby górnicze – Obudowa – Zasady projektowania.
BN-75/0434-05 Wyrobiska komorowe – Obudowa – Obliczenia statyczne i projektowanie.
BN-78/0434-07 Wyrobiska korytarzowe i komorowe w kopalniach – Obudowa powłokowa – Wytyczne projektowania i obliczeń statycznych.
BN-79/0434-03 Szyby górnicze – Obudowa – Zasady projektowania.
BN-79/0434-04 Wyrobiska korytarzowe w kopalniach – Obudowa sklepiona – Wytyczne projektowania i obliczeń statycznych.
BN-83/0434-02 Szyby górnicze – Obudowa – Obciążenia.
Borecki M. (et al.) 1959: Poradnik górnika, vol. II, part 1, Katowice, WGH.
Chudek M. (et al.) 2013: Nowe wyzwania i metody w projektowaniu głębienia i pogłębiania szybów podstawą rozwoju górnictwa w Polsce. Zagadnienia wybrane. Monografia. Gliwice, Katedra Geomechaniki i Budownictwa Podziemnego i Zarządzania Ochroną Powierzchni Wydziału Górnictwa i Geologii Politechniki Śląskiej w Gliwicach.
Czaja P. 2002: Analiza nośności segmentowej obudowy szybów upodatnionej materiałem nieliniowo sprężystym. Kraków, Wyd. Instytutu Gospodarki Surowcami Mineralnymi i Energią PAN.
Domańska D. 2003: Nowa metoda szacowania nośności odrzwi obudowy stalowej łukowej oparta na teorii nośności granicznej. Archiwum Górnictwa, 1.
Duży S. 2007: Studium niezawodności konstrukcji obudowy i stateczności wyrobisk korytarzowych w kopalniach węgla kamiennego z uwzględnieniem niepewności informacji. Zeszyty Naukowe Politechniki Śląskiej, nr 1750.
Kostrz J. 1964: Głębienie szybów specjalnymi metodami. Katowice, Wyd. Śląsk.
Krupennikov G.A., Bulyčev N.S., Kozel A.M., Filatov N.A. 1966: V zaimodejstvie massivov gornych porod s krepju vertikalnych vyrabotok. Moskva, Izd. Nedra.
Mohr F. 1946: Grundlagen der Berechnung des Ausbaues von Schächten unter besonderer Berücksichtigung von Gefrierschächten. Bergbau – Archiv, 2.
Mohr F. 1950: Die Beanspruchungen und Berechnungen des Schachtausbaus. Glückauf, 23/24.
Murzewski J. 1989: Niezawodność konstrukcji inżynierskich. Warszawa, Wyd. Arkady.
Nguyen H.B., Wichur A. 1985: Model losowej zmienności wytrzymałości betonu przy ściskaniu mimośrodowym. Archiwum Górnictwa, 2.
Olszewski J., Kostrz J.A. 1995: Udział Przedsiębiorstwa Budowy Szybów S.A. w rozwoju budownictwa szybowego w Polsce. Sympozjum z okazji 50-lecia Przedsiębiorstwa Budowy Szybów w Bytomiu. Bytom, Przedsiębiorstwo Budowy Szybów S.A.
Pękacki W., Rycman S., Tokarz A. 1996: Polskie budownictwo górnicze – doświadczenia i przyszłość. Budownictwo Górnicze i Tunelowe, 2.

PN-64/B-03001 Konstrukcje i podłoża budowli – Zasady projektowania i obliczeń statycznych.

PN-76/B-03001 Konstrukcje i podłoża budowli – Ogólne zasady obliczeń.

PN-84/B-03264 Konstrukcje betonowe, żelbetowe i sprężone. Obliczenia statyczne i projektowanie.

PN-B-03264:2002 Konstrukcje betonowe, żelbetowe i sprężone – Obliczenia statyczne i projektowanie.

PN-G-05015:1997 Szyby górnicze – Obudowa – Zasady projektowania.

PN-G-05016:1997 Szyby górnicze – Obudowa – Obciążenia.

Rułka K., Wojtusiak A., Pękacki W., Stochel D. 1996: Doświadczenia krajowego zaplecza projektowo-badawczego i przedsiębiorstw w realizacji wyrobisk tunelowych i budowli podziemnych metodami górniczymi. Katowice, Główny Instytut Górnictwa.

Šejnin V.I., Ruppenejt K.V. 1969: Nekotorye statističeskie zadači rasčeta podzemnych sooruženij. Moskwa.

Smirnow N.W., Dunin-Barkowski I.W. 1966: Krótki kurs statystyki matematycznej dla zastosowań technicznych. Warszawa, PWN.

Szczepankiewicz E. 1985: Zastosowania pól losowych. Warszawa, PWN.

Tajduś A., Wichur A. 2009: Budownictwo podziemne w Polsce – nauka i praktyka (na 90-lecie powstania AGH). Przegląd Górniczy, 5–6.

Ustawa z dnia 3 kwietnia 1993 r. o normalizacji (Dz. U. RP Nr 55, poz. 251).

Walewski J. 1965: Projektowanie szybów i szybików. Katowice, Wyd. Śląsk.

Wichur A. 1970a: Analiza matematyczna obserwacji i materiałów z pomiarów odkształceń obudowy szybów w warunkach ROW i LGOM. Zakład Badań i Doświadczeń Budownictwa Górniczego, temat nr 7/66, etap XIV, Mysłowice, listopad 1970 (praca niepublikowana).

Wichur A. 1970b: Ciśnienie górotworu na obwodzie obudowy szybu jako stacjonarna funkcja losowa. Archiwum Górnictwa, 1.

Wichur A. 1970c: Pole naprężeń w sprężystym pierścieniu kolistym obciążonym stacjonarną funkcją losową. Archiwum Górnictwa, 2.

Wichur A. 1971a: Możliwość wykorzystania do zagadnień projektowych matematycznego modelu ciśnienia górotworu na obudowę szybu jako funkcji losowej. Projekty – Problemy, 10.

Wichur A. 1971b: Wytężenie sprężystego pierścienia kolistego obciążonego stacjonarną funkcją losową. Archiwum Górnictwa, 1.

Wichur A. 1971c: Z prac nad nową metodyką obliczania obudowy szybowej (część I). Budownictwo Górnicze, 2.

Wichur A. 1971d: Z prac nad nową metodyką obliczania obudowy szybowej (część II). Budownictwo Górnicze, 3.

Wichur A. 1972a: Możliwość prognozowania kształtu i wartości obciążenia obudowy szybu w zagadnieniach projektowych przy wykorzystaniu modelu probabilistycznego. Zeszyty Naukowe AGH, 374, zeszyt specjalny 36, Kraków.

Wichur A. 1972b: Nowa metodyka obliczania obudowy szybowej w świetle opracowanych nowych norm branżowych. Budownictwo Górnicze, 2.

Wichur A. 1972c: Prawdopodobieństwo przekroczenia wytrzymałości materiału sprężystego pierścienia kolistego obciążonego stacjonarną funkcją losową. Archiwum Górnictwa, 2,

Wichur A. 1972d: Z prac nad nową metodyką obliczania obudowy szybowej (część III). Budownictwo Górnicze, 1.

Wichur A. 1976: Dwuwymiarowy model obciążenia obudowy szybu jako wektorowej funkcji losowej. Zeszyty Naukowe AGH, 584, Górnictwo 88, Kraków.

Wichur A. 1986: Nowe poglądy na temat obliczania obciążeń obudowy szybowej. PAN Oddz. w Krakowie, Prace Komisji Górniczo-Geodezyjnej, Górnictwo 24, Wybrane zagadnienia z budownictwa podziemnego, Wrocław – Warszawa – Kraków – Gdańsk – Łódź.

Wichur A. 1996: Nowe normy projektowania obudowy szybów górniczych. Budownictwo Górnicze i Tunelowe, 4.

Wichur A., Krywult J., Stąpor J., Domańska D. 1991: Zasady projektowania obudowy wstępnej szybów głębionych z użyciem metody zamrażania górotworu. Konferencja naukowo-techniczna: Budownictwo górnicze i podziemne w nowych warunkach gospodarowania. OBR BG Budokop w Mysłowicach, Kokotek k/Lublińca, 16–17.IX.1991.

Wichur A., Żyliński R. 2000: Underground Construction in Poland – Achievements and Perspectives of Development. Konferencja Nauk.-Techn.: Budownictwo podziemne 2000. Kraków, UWND, AGH, 25–27 września 2000.

International Mining Forum 2015, Kicki et. al. (eds) © 2015 Taylor & Francis Group, London, UK. ISBN 978-1-138-02820-3

The Construction of Shafts in the Copper Fields in the Conditions of the Fore-Sudetic Monocline as Exemplified by the SW-4 Shaft

Wojciech Chojnacki
PeBeKa S.A.

Sławomir Fabich
KGHM Cuprum Sp. z o.o.

Tadeusz Rutkowski
PeBeKa S.A.

1. INTRODUCTION

For 55 years the company Przedsiębiorstwo Budowy Kopalń PeBeKa S.A. with its registered office in Lubin has been providing specialist services in the field of mining and engineering construction based on state-of-the-art technologies. Since its establishment the company has sunk numerous shafts with a combined length of 29 kilometres in the conditions of the Pre-Sudetic Monocline. The geological and hydro-geological conditions of the Monocline result in the necessity of the specialist preparation of the rock mass for shaft sinking operations by the freezing of the strongly waterlogged, loose and cohesive Caenozoic formations as well as the cohesive and considerably waterlogged upper and middle variegated sandstone formations. The very shaft sinking operations have been carried out by means of the mechanical rock mining as well as the drill and blast technology. This article presents the shaft sinking technologies used during the construction of the SW-4 shaft. Over many years these technologies have been developed and adjusted to the specific properties of the rock mass around the shafts. At this point in time they have been verified as effective and guaranteeing complete safety of conducted operations.

2. GEOLOGICAL AND HYDRO-GEOLOGICAL CONDITIONS

The area of the SW-4 shaft is located at the edge of the copper ore deposit mined by the "Polkowice-Sieroszowice" Copper Mine and to the south of the "Głogów Głęboki Przemysłowy" copper ore deposit. The geological profile of the SW-4 shaft can be divided into two main stratigraphic complexes: a complex of Caenozoic sediments constituting a cover of the monocline and a complex of Permian and Triassic sedimentary rocks making up the Pre-Sudetic Monocline (Fig. 1).

The Caenozoic sediments can be divided into Quaternary formations as well as Neogene and Paleogene formations. In the area of the SW-4 shaft, the bottom of the Quarternary formations occurs at a depth of approximately 57.2 m bgl, while the bottom of the Neogene formations at a depth of approximately 357.05 m bgl. The bottom of the Caenozoic formations, which is simultaneously the bottom of the Paleogene formations, lies at a depth of 40815 m bgl.

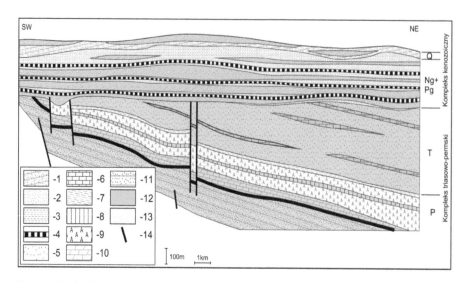

Figure 1. The hydrogeological structure of the area of the Pre-Sudetic Monocline:
1 – clays, 2 – sands, 3 – silts, 4 – lignite, 5 – Triassic sandstones, 6 – limestones
(oolitic, sandy), 7 – mudstones, 8 – dolomites, 9 – anhydrites, 10 – limestones,
11 – Permian sandstones, 12 – aquifers, 13 – insulation layers, 14 – faults

In the Neogene profile, we can distinguish two stages of sediments:

Pliocene – makes up a complex of sandy and silty sediments with a thickness of approximately 116 m, characterised by exceptionally restless sedimentation. Silts are variegated in colour; sands are of various grain sizes up to gravelly.

Miocene – occurs in a cohesive complex of sediments over the whole area around the shaft, reaching a thickness of approximately 184 m. It is divided into three stages in which sedimentation ends with a lignite deposit. The Miocene deposits comprise quartz sands, rarely gravels, mudstones and silts.

The Paleogene deposits are divided into two stages:

Oligocene – occurs in a continuous manner over the whole area in the form of light brown silts, dark grey and fine-grained sandstones as well as grey and green slimes with a combined thickness of approximately 51.0 m. In their roof there occur a few layers of lignite forming the so-called Głogów deposit. The Oligocene layers include sandy layers with a thickness of approximately 44.0 m, constituting a significant water reservoir.

Eocene – occurs in the direct vicinity of the SW-4 shaft in the form of dark grey and brown silty slime with a thickness of approximately 0.5 m, lying directly on the Triassic sandstone.

The Mesozoic sediments in the form of a complex of Triassic rocks lie directly under the Caenozoic sediments. They are represented exclusively by the Lower Trias – variegated sandstone, comprising sandstone rocks of the lower and middle variegated sandstone.

The middle variegated sandstone with the bottom at a depth of 665.55 consists of fine-grained and medium-grained sandstones with interbeddings of mudstones and silty shales. These sandstones are at places fractured and brecciated.

The lower variegated sandstone is made of fine-crystalline, interbedding and brick red quartz sandstones. They are accompanied by a stratum of brown silt shale with a thickness of approximately 11 m. Under this stratum there occur alternately the strata of grey quartz sandstone separated with the banks of brick red sandstone.

The Permian strata are formed as marine Permian limestone lying atop the land formations of new red sandstone. The Permian limestone formations have the form of the sediments of four sedimentation cyclothems. The cyclothems comprise the lithological horizon of anhydrite, base limestone, main dolomite, rock salt and copper-bearing shale.

In the area of the SW-4 shaft, new red sandstone occurs only in the zone of the foundation of the shaft bottom. Generally, new red sandstone is formed as a complex of land sediments with a thickness of approximately 300 m. Its lower part is made up of alternate strata of conglomerates, greywacke sandstones and silty shales. Its upper part constitutes a monotonous series of brown-red and red sandstone, gritstone in the bottom. The roof layers of the new red sandstone slope in northeasterly direction at a slight angle, in accordance with the regional dip in this part of the Pre-Sudetic Monocline.

3. THE FREEZING OF THE ROCK MASS

The freezing of the rock mass necessary for the sinking of the SW-4 shaft was carried out with the application of diversified depths of the freezing holes which were drilled to a depth of 430 m (short holes) and 655 m (long holes). The freezing holes were located at the freezing circle with a diameter of 16 m, and the inspection and measurement circles were located outside the circle, along a common radius and at respective distances of 2, 3 and 5 m from the circle of the freezing holes.

The nominal distances between the freezing holes were as follows:
- up to a depth of 430.0 m – 1.25 m,
- at a depth of from 430 to 655.0 m – 2.50 m.

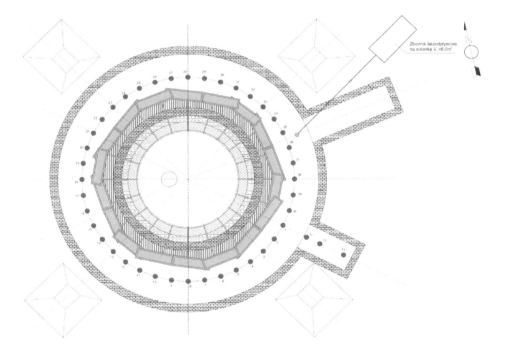

Figure 2. A circle of the freezing holes at the SW-4 shaft

The layout of the freezing holes is presented in Figure 2.

In view of the company's extensive experience, in order to maintain verticality within the limits of the technical requirements, the drilling of the freezing holes was carried out with the application of the RVDS system. This made it possible to minimise the holes' deviations from the perpendicular, and thus to build a uniform and more hermetic ice wall around the shaft.

The freezing process was commenced at the end of 2007. It was conducted in two phases. The first phase consisted in increasing the intensity of freezing so that the ice wall would achieve the thickness specified in the project design. The second phase consisted in maintaining the intensity of freezing in order to keep up the achieved parameters of the ice wall. The operating parameters of the freezing station as well as the growth and stability of the ice wall were monitored on a regular basis. The freezing station was switched off in the middle of 2010. It was determined by the shaft bottom's achievement of the depth of 653.1 m and the installation of the tubbing support down to that depth with the final support ring no. 428 (type N120S). Furthermore, the freezing station remained in the standby mode until the moment of the installation of the final tubbing ring at the depth of 668.71 m and its crowning with basic curb no. 1.

4. THE SHAFT SINKING TECHNOLOGIES USED
   ## IN THE CONSTRUCTION OF THE SW-4 SHAFT

The technologies used in the sinking of the SW-4 shaft were determined by the conditions related to the preparation of the rock mass for the sinking operations, the type of rock, the degree of the flooding of the rock mass and the type of the used support. In the initial stage of the shaft sinking operations, the diaphragm walls technology was applied to construct the shift collar. The company had used this technology earlier in the construction of underground facilities as well as some stations and tunnels of the Warsaw metro. The application of this technology allowed the company to construct the shaft collar without the necessity of using other facilities, which increased the safety of the conducted operations.

In the flooded rock mass (Caenozoic formations and middle variegated sandstone) which was subjected to freezing, a tubbing support with a concrete cladding was used; below (lower variegated sandstone, Permian limestone and new red sandstone) the company used monolithic concrete, reinforced concrete or concrete with drainage at water-logged sections. In the locations where the profile included a stratum of rock salt, a multilayer lining strengthened with a steel, circular and yielding support was used. In the strongly flooded zone, below the freezing range, the company used a short section of a tubbing support again.

To summarise, in the process of the sinking of the SW-4 shaft, the following technologies were used:

Technology I – the sinking of the shaft in a tubbing support in the frozen rock mass divided into:

I.1. the sinking of the shaft in a tubbing support in the frozen rock mass – the breaking of the rock mass by means of the KDS-2 shaft tunnelling machine;

I.2. the sinking of the shaft in a tubbing support in the frozen rock mass – the breaking of the rock mass by means of explosives.

Technology II – the sinking of the shaft below the frozen zone divided into:

II.1. the sinking of the shaft in a concrete and reinforced concrete support with the application of slipform shutter – the breaking of the rock mass by means of explosives;

II.2. the sinking of the shaft in a tubbing support – the breaking of the rock mass by means of explosives;

II.3. the sinking of the shaft in a concrete support with a rock mass drainage system – the breaking of the rock mass by means of explosives;

II.4. the sinking of the shaft in a rock salt layer in a multilayer lining strengthened with a steel, circular and yielding support – the breaking of the rock mass by means of the KDS-2 shaft tunnelling machine.

4.1. *The technical support facilities used in the sinking of the SW-4 shaft*

The site layout plan for the period of the shaft sinking operations provided for the construction of the following facilities (Fig. 3, 4):
– a mine headgear,
– sinking hoist buildings,
– a shed for shaft winches,
– winches and hoists,
– a freezing room and installation,
– temporary electrical buildings,
– a mechanical and electrical workshop,
– a pneumatic equipment repair and maintenance workshop,
– a compressor station and a compressed air installation,
– a receiving and temporary material storage bin,
– a tubbing storage yard,
– a main transformer station,
– a shaft fan and a fan building.

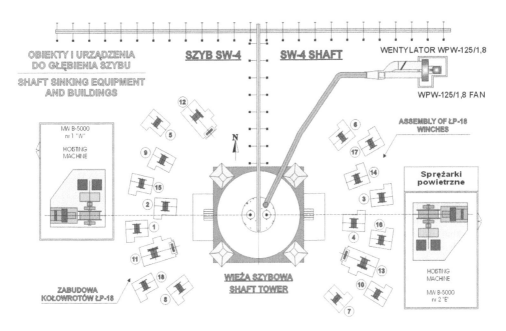

Figure 3. The site layout plan for the period of the shaft sinking operations

Figure 4. The shaft headgear for the period of the shaft sinking operations

4.2. *The shaft equipment for the period of the shaft sinking operations*

For all technologies applied, the equipment of the SW-4 shaft during the sinking operations was similar. The basic elements of the equipment comprised the following:
– a three-landing working platform (Fig. 5),
– a shaft loader,
– a two-landing platform for cementation operations.
  The machinery used depending on the applied shaft sinking technology:
– a shaft drilling rig,
– a shaft sinking machine,
– a tubbing support installation ring,
– hydraulic formwork.

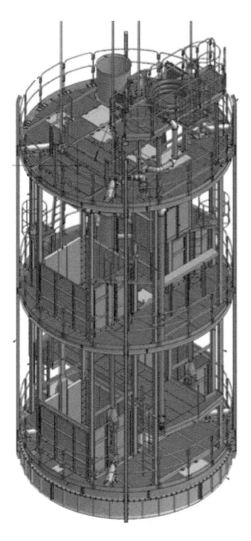

Figure 5. A three-landing working platform

4.3. *Technology I – the sinking of the shaft in a tubbing support in the frozen rock mass*

In the technologies I.1 and I.2, the installation of the tubbing rings was carried out by means of an installation ring propped at the shaft bottom with hydraulic advancing cylinders and guided on the lines of three slow-speed drum winches located in the shaft site.

4.3.1. *The sinking of the shaft in a tubbing support in the frozen rock mass – the breaking of the rock mass by means of the KDS-2 shaft tunnelling machine*

In the zone of the mining operations conducted in the originally loose and flooded, and subsequently frozen, rock mass (Caenozoic formations), the installation ring cooperated with the KDS-2 shaft sinking machine (Fig. 6, 7).

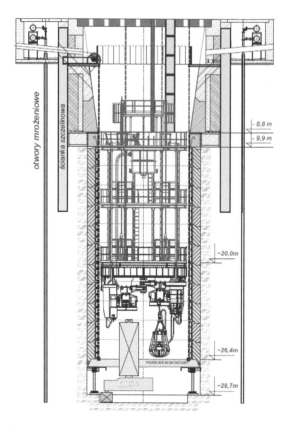

Figure 6. The equipment of the shaft during the sinking operations
by means of the shaft sinking machine in the tubbing support

Figure 7. The head of the shaft sinking machine

Figure 8 presents the course of the whole technological cycle comprising the breaking of the rock mass by means of the KDS-2 shaft sinking machine in a breakout with a diameter of 9.22 m and the installation of one tubbing ring with a height $H = 1.5$ m.

4.3.2. The sinking of the shaft in a tubbing support in the frozen rock mass – the breaking of the rock mass by means of explosives

After the shaft bottom has reached the zone of the roof of frozen cohesive rocks, the ring and the shaft sinking machine are dismantled and replaced with an installation ring adjusted to the performance of blasting operations. The further shaft sinking operations in the tubbing support are carried out with the application of explosives.

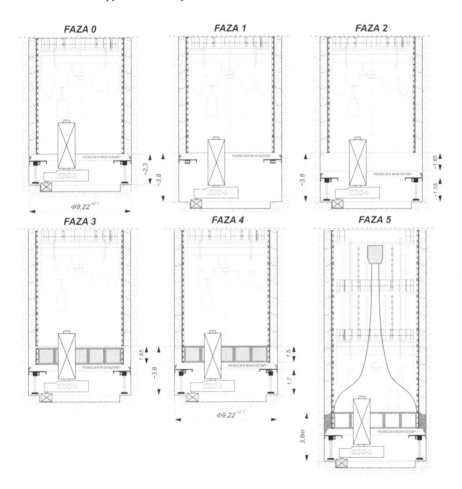

Figure 8. The technological cycle of the construction of a section of the tubbing support
with the breaking of the rock mass by means of the KDS-2 shaft sinking machine.
Phase 0 – the initial state, *Phase 1* – the breaking of the rock mass by means of the KDS machine,
Phase 2 – the sinking and stabilisation of an installation ring, *Phase 3* – the mounting of the tubbing
ring on the installation ring, *Phase 4* – the attachment of the tubbing ring to the existing tubbing column,
Phase 5 – the sinking of concrete behind the tubbing ring

123

Figure 10 presents the technological phases of the full cycle of the shaft sinking operations in the tubbing support at the section of frozen cohesive rocks. It comprises blasting operations whose objective is to make a breakout with a diameter of 9.3 m for the installation of one tubbing ring with a height of 1.5 m.

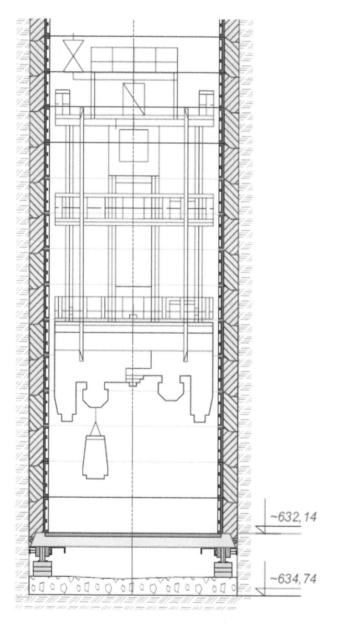

~632,14

~634,74

Figure 9. The equipment of the shaft during the sinking operations
in the tubbing support with the application of explosives

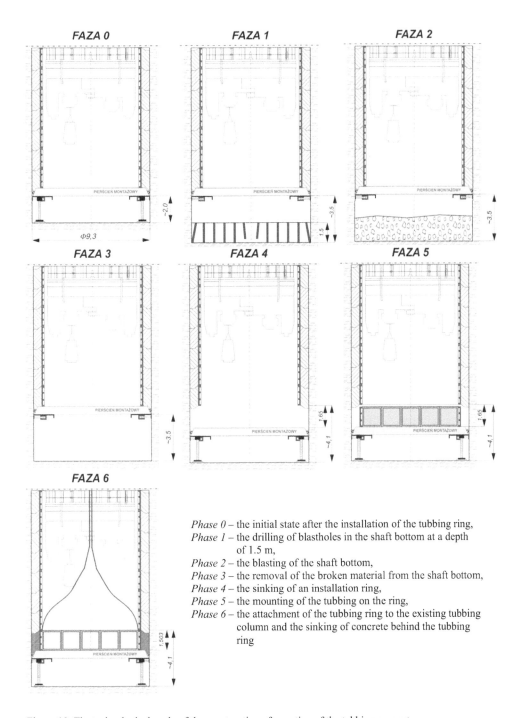

Figure 10. The technological cycle of the construction of a section of the tubbing support
with the breaking of the rock mass by means of explosives

Phase 0 – the initial state after the installation of the tubbing ring,
Phase 1 – the drilling of blastholes in the shaft bottom at a depth
  of 1.5 m,
Phase 2 – the blasting of the shaft bottom,
Phase 3 – the removal of the broken material from the shaft bottom,
Phase 4 – the sinking of an installation ring,
Phase 5 – the mounting of the tubbing on the ring,
Phase 6 – the attachment of the tubbing ring to the existing tubbing
  column and the sinking of concrete behind the tubbing
  ring

4.4. *Shaft sinking technology II – the shaft sinking operations below the frozen zone*

After the bottom of the frozen zone has been reached there occurs a change in the types of the shaft support and consequently a partial change in the shaft's equipment necessary for the construction of the support. The tubbing installation ring is dismantled and replaced by steel hydraulically propped timbering hung on hoist lines (Fig. 11), which allows the construction of a concrete support.

Blastholes below the frozen zone are bored by means of hand-held pneumatic drills and a shaft drill rig (Fig. 12).

Figure 11. Slipform shutter for the SW-4 shaft

Figure 12. A shaft drill rig

4.4.1. *The sinking of the shaft in the concrete and reinforced concrete support with the application of slipform shutter*

The figure below presents the course of the whole technological cycle of the sinking of the shaft together with the breaking of the rock mass by means of the blasting technique for the purpose of the installation of one segment of the concrete support with a height $H = 3.5$ m with the application of a hydraulic steel hanging timbering system.

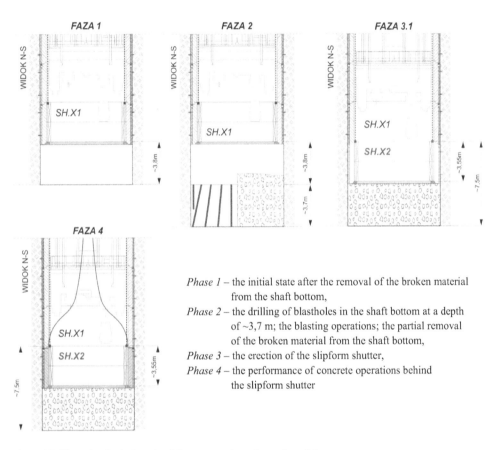

Phase 1 – the initial state after the removal of the broken material from the shaft bottom,

Phase 2 – the drilling of blastholes in the shaft bottom at a depth of ~3,7 m; the blasting operations; the partial removal of the broken material from the shaft bottom,

Phase 3 – the erection of the slipform shutter,

Phase 4 – the performance of concrete operations behind the slipform shutter

Figure 13. The technological cycle of the construction of a section of the concrete support with the breaking of the rock mass by means of explosives

4.4.2. *The sinking of the shaft in the tubbing support – the breaking of the rock mass by means of explosives*

The flooded zone of the shaft, below the frozen zone and covering the interval of 835÷864 m, was closed with a tubbing support allowing the transfer of the whole range of occurring loads. The support consisted of 34 tubbing rings installed at a depth range of from 823.37 to 874.53 m. The equipment of the shaft was the same as in technology I.2.

4.4.3. *The sinking of the shaft in the concrete support with the drainage of the rock mass*

In the SW-4 shaft, at a depth range of from 958.36 to 985.73 there occurs a stratum of main dolomite closed from the top by a cohesive dolomite and anhydrite breccia, and from the bottom – by a package of cohesive anhydrites. This section of the shaft was equipped with a concrete support with the drainage of water and its evacuation through a collector. At this section, the shaft was sank in a preliminary support. After the installation of the basic curb and the elements of the drainage system the concrete support was constructed in the "from bottom to top" system with the application of alternate formwork. The water coming from the shaft cheek flowed over a drainage matt to the drainage collector. The drainage matt was laid on the shaft cheek in horizontal strips. In order to minimise the penetration of the concrete support with water and to ensure its leaktightness the whole surface of the drainage matt was sprayed with the Krzemopur ES and Krzemopur systems. The drainage system collector was installed on the basic curb (Fig. 14). From the upper part of the collector, at regular intervals along the shaft's circumference, 6 pipes (PE DN 50) led out towards the shaft's interior. The pipes' sections running in the sidefill are perforated. Their task is to prevent increases of water pressure in the collector in the event of a blockage in the collector pipe.

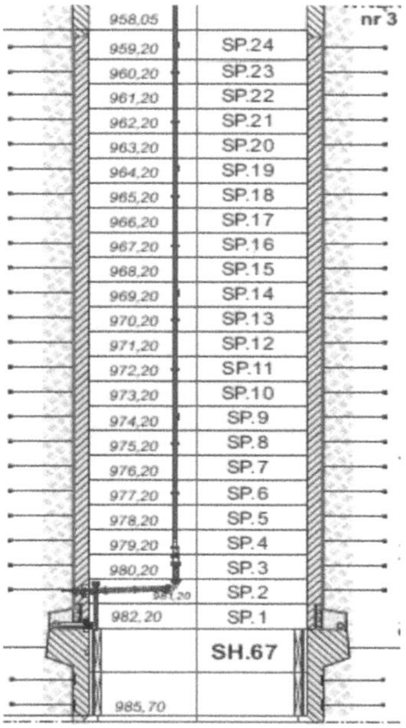

Figure 14. The construction of the final support in the dolomite stratum with the application of alternate formwork

Figure 15. The construction of the drainage collector in the dolomite stratum

4.4.4. *The sinking of the shaft in a rock salt stratum in a multilayer lining strengthened with a steel, circular and yielding support*

In the SW-4 shaft, at a depth range of from 1026.36 to 1185.49 m there occurs a rock salt stratum. Its thickness of 152.65 m is the largest of all such strata identified within the profiles of the LGOM shafts. It is also located at the deepest level of all such rock salt strata. In view of the rheologic properties of rock salt and very large pressure of the rock mass, for both technical and economic reasons, a decision was made not to install a support with a full load-bearing capacity, but to replace it with a multilayer lining support (Fig. 15). As a multilayer lining protecting the salt wall against the atmosphere of the shaft, a system applied in spray and consisting of a polyurethane prime coat Ekopur LS/G, a plastic lagging net and a proper membrane protecting the side wall called Krzemopur was used. After the application of the membrane the salt wall was secured additionally by means of a steel flexible arch support. This measure is to protect the shaft working against the consequences of the dilatancy process. The breaking of rock salt was carried out by means of the same shaft machine as that used in the breaking of the frozen rocks of the Caenozoic formations.

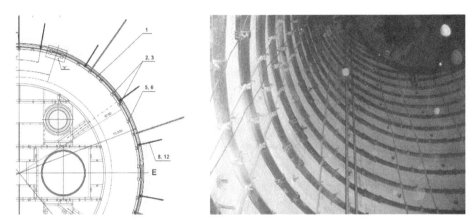

Figure 16. A multilayer lining support strengthened with steel arches in a rock salt stratum

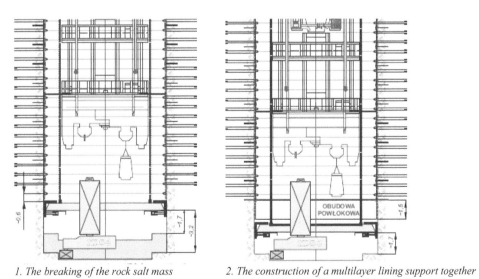

1. The breaking of the rock salt mass

2. The construction of a multilayer lining support together with the application of a net and the installation of rings

Figure 17. The performance of the technological cycle related to the sinking of the shaft in the rock salt layer, in the multiple lining support, with the application of the shaft sinking machine

SUMMARY

The shaft sinking technology currently used in LGOM is adjusted to the specific geological and hydrogeological conditions of the copper ore deposits and based on over 40 years of operating experience. With respect to the safety of the shaft sinking operations, this technology has been confirmed as guaranteeing the timely performance and complete safety of planned works. The occurring problems result more from the changeability of the geological and mining conditions as well as a relatively poor reconnaissance of the local hydrodynamic conditions despite the application of the latest research methods. The additional methods of examining the rock mass based on pilot holes drilled during shaft sinking operations are characterised by a high level of effectiveness, which facilitates correct choices of technologies.

It should be emphasised that the technologies presented in this article have their weaknesses – mainly with respect to the effectiveness of shaft sinking operations conducted in the cohesive rock mass.

However, it is difficult to compare the pace of such operations conducted at various locations around the world because there is no common plain of reference, mainly with respect to the geological conditions occurring in the area of the Pre-Sudetic Monocline.

Various attempts have been made to increase the effectiveness of shaft sinking operations. They consist in the following:
- the introduction of unconventional methods of driving openings and constructing supports to the underground construction practices in the conditions of copper ore deposits;
- the optimisation of the processes of rock mass preparation for drivage operations with respect to both the selection of proper methods and their technological parameters;
- changes in work organisation leading to the elimination of activities which are not directly related to the shaft sinking rate as well as the reduction of time consuming and expensive operations or actions.

International Mining Forum 2015, Kicki et. al. (eds) © 2015 Taylor & Francis Group, London, UK. ISBN 978-1-138-02820-3

Underground Deposit Exploitation – Bontang Project

Janusz Olszewski, Paweł Wójcik, Janusz Rusek
KOPEX – Przedsiębiorstwo Budowy Szybów S.A.
(KOPEX – Shaft Sinking Company S.A.)

ABSTRACT: In 2008, KOPEX Group made a decision to enter Indonesia's mining market. Indominco Stage 1 Underground Mine was the first Mining Project carried out in Bontang, Indonesia.

Our works related to the construction of two declines and trial coal extraction from C13 seam allowed us for gathering new expertise in the field of conducting mining operations in a new market and in room and pillar system, unprecedented in the Polish coal mining.

1. INTRODUCTION

From the very beginning, KOPEX Group took steps aimed at expanding into new global markets for the supply of mining machinery, equipment and services.

KOPEX Group decided to enter the Indonesia's market with comprehensive coal mining services and equipment, considering rise in demand for coal in the world, increase in coal production in Indonesia as well as scheduled cease to open-cast mining in Indonesia in the future.

In May 2008, the KOPEX Group's companies: KOPEX S.A. and KOPEX – Przedsiębiorstwo Budowy Szybów S.A. (hereinafter referred to as KOPEX – PBSz) signed a contract for executing underground mining operations for the PT Indominco Mandiri which formerly conducted coal open-cast mining in the framework of Bontang Project Stage I.

The investor planned to develop this underground trial mine in two stages. The ultimate goal of the project was construction of coal mine in which pillar mining system would be applied and annual target production volume amounted to more than 1 Mt.

The mine expansion was carried out in two stages, as follows:

STAGE I – expansion of trial mine galleries and run of the first panel extracted in room-and-pillar system.

Coal panel development will be followed by successive pillar mining. It will be commenced with a relatively low mining rate percentage extraction and continued with its gradual progress to the full scheduled mining rate.

Stage I was aimed at determining what percentage amount of coal can be safely extracted to finalize the most appropriate engineering design for mine and mining systems in Stage II.

STAGE II – after a successful completion of Stage I, the mine can be further expanded using the design parameters determined in Stage I, to increase coal production capacity to more than 1 Mt per year.

Stage I mining work started directly from the end location of the completed "trial adit".

The trial adit was driven from the opencast pit stopped wall, opening C13 Indominco seam, in the neighbourhood of the existing surface equipment and mine waste dump.

In 2008, PT Kopex Mining Contractors (KMC), based in Jakarta was set-up by the KOPEX group's companies for effective operate in the Indonesia's market.

The KMC, the newly established legal entity, incorporated under Indonesian law, took over conducting the contract on construction of the Bontang Coal Mine for PT Indominco Mandiri.

Stage I was completed in February 2010, after taking over full responsibility for conducting the contract by the KMC in August 2009.

2. DEPOSIT MINING – BONTANG PROJECT

2.1. *Indonesian Mining*

Indonesia's market has always been seen as a promising market. The first attempts of coal mining on an industrial scale in Indonesia date back to the late 19th and early 20th century.

In 2002, Indonesia produced more than 100 Mt of coal, of which nearly 73 Mt went to foreign markets. Indonesia's hard coal mining is dominated by private investors: PT. Adraro Indonesia, PT. Arutmin Indonesia and PT. Kaltim Prima Coal. PT. Tambang Batubara Bukit Asam, the only state-owned mine, produces coal in the amount of ca 10 Mt per year (production in 2002 amounted to 9,482,041 t).

Indonesia is one of the leading producers of coal in the world. According to the Ministry of Energy and Natural Resources of the Republic of Indonesia, seams located in Indonesia are estimated at nearly 40 Bt of coal of which11.5 Bt have already been explored and classified, and the remaining resources are only estimates. Commercial coal reserves are estimated at nearly 6 Bt. Major hard coal deposits are located in Sumatra (nearly 2/3 of the total resources), Kalimantan, West Java and Sulawesi. In Sumatra, the largest coal deposits are located in the vicinity of Tanjung Enim and operated by PT. Tambang Batubara Bukit Asam, a state mine. Coal of best commercial value is deposited in Kalimantan is mined by private contractors and 95% of coal deposits are extracted by opencast method.

Underground mining method, which allows for production of better quality coal has not been used large- scale so far. Easily accessible coal seams extracted by opencast method slowly comes to an end. Coal producers are beginning to consider extracting deeper deposited and hard accessible coal seams using underground mining method. Underground mining conditions in Indonesia are very difficult, due to high seam inclinations and rock water-sensitivity but coal deposited underground is characterized by much better quality parameters.

Calorific value of the Indonesian coal is in the range of 5,000 to 7,000 kcal/kg and it is characterized by low ash and sulphur and high moisture contents.

From commercial point of view, the Indonesian coal is similar to coal quality offered by the Republic of South Africa. Lignite coal, covering about 59% of total Indonesia's coal resources, is used only in the domestic market. Its export is not profitable, due to low calorific value and high moisture content (over 30%). The remaining part of coal is exported to the neighbouring countries as well as to the United States and Europe. The average sulphur content in the Indonesian coal amounts to 1.0%.

2.2. *BONTANG project*

The Bontang Coal Mine is located in Borneo, in East Kalimantan Province, in PT Indominco Mandiri's mining area.

The mine is situated on the surface of the closed open cast mine, in the best place to conduct underground mining. It is the central part of PT Indominco's west concession block, of the area of ca 50 km^2 (10×5 km).

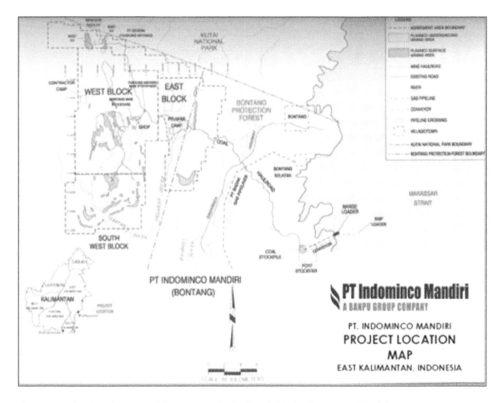

Figure 1. Project location map of the Bontang Coal Mine. PT Indominco Mandiri mining area

2.3. *Deposit description*

There were determined the following deposit parameters based on geological explorations and data obtained from geological interpretation:
- deposition depth of every seam roof,
- coal seam thicknesses,
- coal seam identification (correlation).

For its most part, the deposit is formed in a wide syncline in north- south direction and slightly falling towards the south.

Coal seams fall at an angle of 10°, maximum 12° in most cases. Seam thicknesses vary from a few cm to 9 m.

The C4, C6, C7, C8, C11, L11 and C13 seams are characterized by an average thickness of more than 1 m.

From among all the coal seams deposited in the West Block area, only C13 seam can be mined by underground method. C13 seam thicknesses range from 0.5 m to 9.0 m.

In some areas, C13 seam is eroded. In the west wing of the syncline, this geological phenomenon is particularly evident. One can distinguish around five eroded areas, whose locations were specified by additional drilled holes. There is also one erosion in the east wing, in a distance of about 350 m behind the trial adit.

A thickness of 1.5 m was adopted as seam balance minimum thickness, allowing for carrying out mining operations and a max. permissible interlayer thickness of 0.3 m.

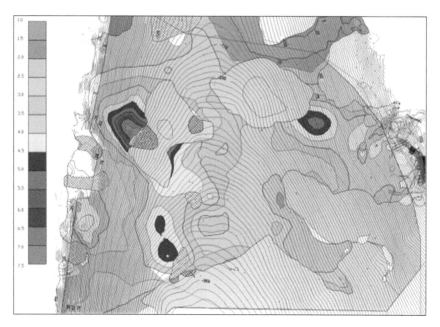

Figure 2. Map of C13 seam thickness with deposition depth isolines

Total proved and indicated reserves with a min. thickness of 1.5 m are estimated at approx. 44.7 Mt, including:
– 36.0 Mt of reserves with a min. thickness of 2.5 m,
– 24.8 Mt of reserves with a min. thickness of 3.0 m,
– 15.4 Mt of reserves with a min. thickness of 3.5 m.

Coal deposited in C13 seam is bright, finely laminated, featured by nearly vertical cleavage facets of dominant east-west orientation.

Coal has been classified as moderately strong, showing a compressive strength of 15 MPa in a UCS test.

In direct roof of C13 seam, there is a layer of mudstone and fine sandstone shoaling, with a thickness between 1 m and 12 m (approx. 6 m). Compressive strength of this layer is poor. It has shown only between 2 and 10 MPa in a UCS test.

Above mudstone, there are layers of sandstone, each 1m to 2m thick. Whereas, on open pit wall is as thick as 10 m in total.

In direct floor of C13 seam, there is a layer of carbonaceous shale and a layer of slightly compact mudstone with fine sandstone shoaling underneath. Mudstone thickness varies between 1 m and 17 m, in average it is about 3m. Its compressive strength amounts to 8 MPa but laboratory tests have shown that mudstone is intensively water-sensitive.

Based on an assessment of the geotechnical analysis, there was made assessment on geotechnical issues related to entries engineering design.

Roof in the Bontang Coal Mine was rated at 25–35, on a 1–100 scale of Coal Mine Roof Rating (CMRR).

Coal in this deposit is characterized by better mechanical properties than roof rock, and therefore stability will depend on leaving a large coal piece in the roof. Generally, mudstone coal roof rating (CMR) amounts to 27 when protected against humidity and 23 when exposed.

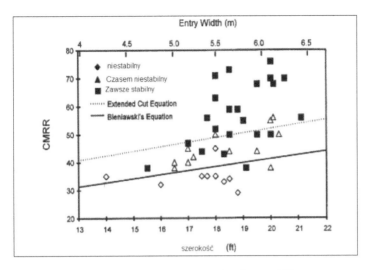

Figure 3. Relation between Roof Rating (CMRR) and entry width

Due to seam inclination, it should be remembered that for a horizontal strike entry, thickness of a large coal piece left, will be different at every of two side walls.

Basing on the above analysis, it was estimated that thanks to a large coal piece left in the roof, props will be used in a limited quantity. In general, long tendon anchors in roof support will be used only in some selected places.

There were designed a few pillars included in the project pre-feasibility study. At a depth of 300 m, the pillar dimensions of the pillar to ensure a satisfactory stability are as follows: 14 m in width (19 m centres) and 20 m in length (25 m centres). To minimize the number of cut through and related four-direction intersections, pillar lengths were extended up to 45 m (50 m centres).

2.4. *Execution of work*

In April 2008 started first work for drilling declines in C13 seam.

Scope of work covered the extension of mine workings from the completed entries of the trial mine to 11 cross gate (Preparatory work) and conducting mining operation in a phase mode system in the south, between 11 and 6 cross gates (Excavation work).

Table 1. Parameters of development workings

Characteristics of development workings	
Length:	2 × ca 2700 m
Width:	4.8 m to 5.5 m
Height:	3.0 m to 3.5 m
Inclination of workings:	8° to 15°
Roof support:	Bolting roof support: 1.5×1.5 m, adhesive bolts of 24 mm in diameter and 2.2 m long + rope bolts 8.0 m long

While conducting mining operation, excavation sequences subjected to continuous revisions based on qualitative observations and changes results. The plan was modified in accordance with the requirements of the Indominco Mandiri Company.

Figure 4. Bontang trail mine area with entries to declines at the old open pit area of C13 seam

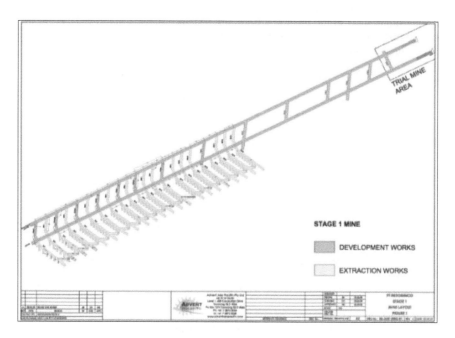

Figure 5. Stage I Bontang mine workings scheme

Maintenance work started when the north and south declines were drilled up to 11 cross gate.

Cross-section dimensions of a standard haulage gate are 5.5 m in width and 2.7 m in height. All haulage gate profiles were designed with a nominal coal layer of approx. 300 mm in roof and floor for protection against poor roof and floor expected and to ensure safe and secure workplace.

Due to specific characteristics of mining workings, the drilling schedule was as follows:
- All pockets were drilled in the distance from to max. 3.5 m from roadheader cutting head.
- On the first 200 m length from 11 cross gate, there was kept a distance of 25 m between the driven galleries. Over the next 200 m, galleries were driven in a distance of 20 m from each other to enhance volume of coal to be mined.
- Splits between 11 and 8A cross gates were scheduled to be made in the linear range. However, due to local conditions, there were implemented necessary modifications in the mining operations plan.
- Ventilation splits were designed for 5.2 m in width, except for the distance of the last 4 m, where the width of ventilation splits was designed for 3.5 m. The remaining 4 m of ventilation split were left with no roof support, in order to protect against the risk of spark igniting between the Continuous Miner cutter head and bolts, during conducting mining operations.
- Workings in the panel were driven upwards at an angle of 60° in relation to the axis of the south decline, at a distance of approx. 28.4m between the maingate and at an angle of 90°, at the junction with the south decline.
- After joining with south decline at an angle of 90°, they were upwards drilled splits 70 m long.
- Extraction pockets were initially drilled at 10m distances from each other for the first two splits and the distances were gradually reduced to 8 m and finally to 6.5 m.
- Plan and dimensions of operational workings were verified at executing works in accordance with monitoring, observation and measurement results of a site.

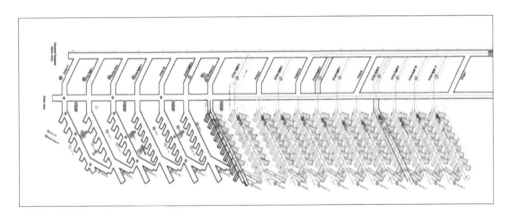

Figure 6. Extraction workings scheme

2.5. Machinery and equipment

Joy 12HM9 Continuous Miner was used to execute work in the first stage. It is a high performance machine manually controlled. It is equipped with a two-drum cutting head and featured by compact structure and centrally located scraper conveyor. Cutting drums are of 1,120 mm in diameter. CM cutting width amounts to 4.9 m and a max. cutting height is 3.75 m. Scraper conveyor is 760 mm wide. Due to weak floor strength parameters, the Joy 12HM9 CM was not suitable for drilling because of its large weight.

ROM coal haulage was affected with JEFFERY 4110 Ramcar tyre vehicles. ROM coal is loaded directly from CM scraper conveyor into the vehicle. Its capacity amounts to 14 t. After loading, ROM coal is directed to the surface.

At a later stage of work, along with successively longer and longer haulage distances. ROM coal haulage was made with electrically powered shuttle cars and main haulage belt conveyor installed in north decline. ROM coal during cutting operation was poured from CM conveyor onto shuttle cars and then transported to main haulage belt conveyor, in a distance of up to 200 m, extended with work progressing.

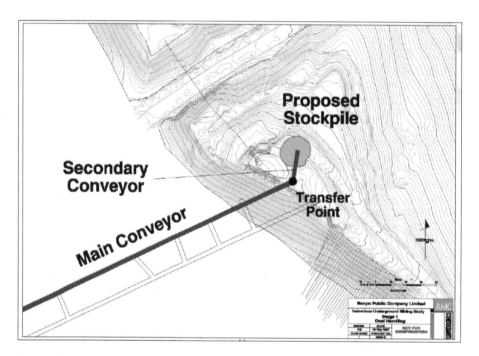

Figure 7. Main conveyor scheme. Belt conveyor installed in north decline

Machinery and equipment used to carry out drilling and of excavation operations are as follows:
– Joy 12 CM 11 Continuous Miner with a roof bolter. There were used only individual roof bolters , due to technical reasons and limited possibilities of simultaneous bolting and mining;
– two Joy 15 SC shuttle cars, electrically powered;
– one ROM coal crusher;
– one PJB Diesel vehicle for personnel transport;
– one set of belt conveyor system, 800 m long;
– stacker belt conveyor on the surface.

2.6. Roof support for mine workings

Bolting roof support with steel mesh was applied in the workings. Steel bolts 2.2 m long were used for bolting, 6 bolts were used for 1 rm of workings.

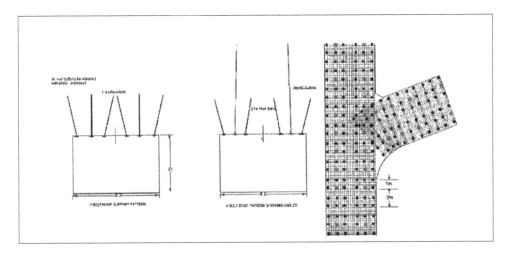

Figure 8. Scheme of roof bolting for Stage I

2.7. *Ventilation system*

Circulating suction ventilation system was applied during drilling operations of declines. Air circulation was forced by a suction fan built-in in the inlet to north decline. Ventilation dams forcing air circulation through the working face were installed in cross-cuts joining declines.

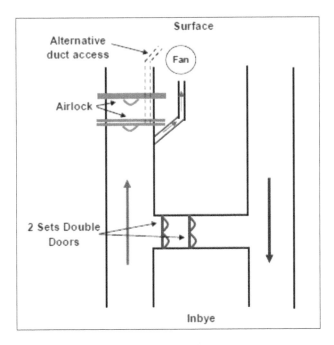

Figure 9. Scheme of ventilation circuit. Entry to declines

2.8. Progresses achieved in drilling

In the initial stage of work, due to logistical problems there was not achieved satisfactory progress of conducting the work.

After taking over the full responsibility for the conduct of the contract in August 2009 by PT KOPEX Mining Contractors, KOPEX Group's company, and changing the organization of supply of materials and spare parts for machinery and equipment, significant drill progresses started to be noted.

Table 2. Monthly driving progress

	Progress in a month [m]	Number of drilling days [days]	Average progress [m/day]	Cumulatively [m]
July 2008 (from 18 July 2008)	60	11	5.5	60
August 2008	84.5	13	6.5	144.5
September 2008	104.2	20	5.2	248.7
October 2008	148.7	22	6.8	397.4
November 2008	252.3	24	10.5	649.7
December 2008	214.8	22	9.8	864.5
January 2009	293.7	28	10.5	1158.2
February 2009	211.5	19	11.1	1369.7
March 2009	120	11	10.9	1489.7
April 2009	171.7	14	12.3	1661.4
May 2009	195.5	15	13.0	1856.9
June 2009	148.4	17	8.7	2005.3
July 2009	0	0	0.0	2005.3
August 2009 (from 22 August 2009)	117.2	10	11.7	2122.5
September 2009	335.1	17	19.7	2457.6
October 2009	723.2	30	24.1	3180.8
November 2009	1300	30	43.3	4480.8
December 2009	812.2	18	45.1	5293
January 2010	1032	29	35.6	6325
February 2010 (until 6 February 2010)	280	14	20.0	6605

Table 3. Summary of driving progress

	Progress [m]	Number of drilling days [days]	Average drilling progress [m/day]
In total	6605	462	14.30
In total until stopping 18 July 2008	2005.3	216	9.28
In total after restarting 22 August 2009	4599.7	246	18.70

SUMMARY

It was a new challenge for the Polish mining and mining companies included in the KOPEX Group to execute work related to opening and conducting trial room and pillar mining operations in C13 seam.

In the initial stage, the projected progress was not achieved due to some difficulties in cooperation with the Australian consortium member.

After taking over the contract and ordering matters relating to the supply of materials and spare parts for machinery and equipment, there was noted a significant increase in the performance achieved at drilling work.

Operations conducted in the Bontang Coal Mine allowed us to gain experience with a new coal mining technology in room-and-pillar system not used in Poland.

Continuous Miner applied in mining operations had to be used because of mining and geological conditions.

The main factors affecting the operation method used are as follows:
- difficult conditions for executing work on water-sensitive floor impede the application of heavy mining machinery;
- high working stability allowing for the use of bolting roof support;
- irregular deposition and irregular seam thicknesses.

Due to the need for using underground mining in Indonesia resulting from the fact that opencast mining will have to be ceased in the future, our activities related to construction of the Bontang Coal Mine and operation of C13 seam was aimed at gathering knowledge and information on the behaviour of the rock mass and capability of applying high-performance longwall systems manufactured by the KOPEX Group to maximize coal production capacity in Indonesia.

REFERENCES

The paper was elaborated on the basis of the KOPEX Group's internal documents and documents related to conducting the contract for PT Indominco Mandiri.

[1] KOPEX S.A.: Engineering Department: Deposit Mining Systems. Katowice 2008.
[2] KOPEX S.A.: Report on Operation of KOPEX S.A.'s Subsidiary in Indonesia. Katowice 2007, 2008.

International Mining Forum 2015, Kicki et. al. (eds) © 2015 Taylor & Francis Group, London, UK. ISBN 978-1-138-02820-3

Deepening the JANINA VI Shaft

Jarosław Jankowski, Jacek Mika, Mariusz Gierat
KOPEX – Przedsiębiorstwo Budowy Szybów S.A.
(KOPEX – Shaft Sinking Company S.A.)

1. PURPOSE OF DEEPENING THE JANINA VI SHAFT

Investing in deepening the Janina VI shaft is of tremendous importance for the existence of the Janina Coal Mine.

Construction of 800 m Level will enable first working implement in a new, IV Mining Level, based on 207 coal seam characterized by good coal quality parameters, regular deposits and high thickness.

In the future, it is going to be the most significant coal seam of the Janina Coal` Mine.

It will allow increasing coal production to 3,600,000 tonnes that will result in increase in employment of about 460 people.

Simulations made have shown that the mine's capability to operate is estimated until 2065.

Mining operation start up in 800m Level is expected in the second half of 2018.

The Janina VI shaft was constructed between 1988 and 1992.

On 25 August 1995 started a gradual flooding of the constructed shaft and its adjacent workings.

KOPEX – Przedsiębiorstwo Budowy Szybów S.A. (KOPEX – Shaft Sinking Company S.A.), hereinafter referred to as KOPEX – PBSz, was commissioned a complete flooding operation of the shaft and the adjacent workings, due to the need of opening out new mining areas.

The works related to deflooding were carried out between 13 July 2005 and 05 August 2006.

In August 2013, there was signed a contract with KOPEX – PBSz for shaft deepening. Its execution was scheduled for 36 months.

2. GENERAL SHAFT CHARACTERISTICS

The Janina VI shaft is of 7.5 m in diameter of and 523 m deep.

Up to the depth of 30 m, the shaft is being constructed in two layers: BSZ-2 precast concrete elements (outer lining) and B15 concrete (inner lining). To the depth of 217 m it is constructed with B15 class concrete and every layer is 40 cm thick. Between the depths of 325 m and 523 m, the shaft is constructed with a double-layer lining; every layer is 40 cm thick.

Currently, the shaft is only providing fresh air to 350 m and 500 m Levels.

The existing section of the shaft is equipped as follows:
– ladder way,
– guides for large-size cage and counterweight, mounted on the furnishing girders, spaced at a distance of 4.5 m,
– shaft chairs at 350 m and 500 m Levels,
– cable cleats with suspended cables,
– compressed air, dripping and drainage pipelines.

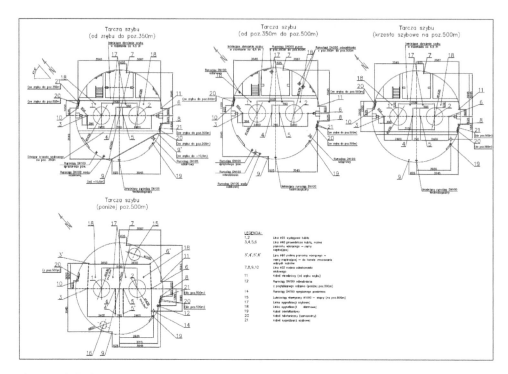

Figure 1. Shaft plans

3. CONCEPTUAL FRAMEWORK FOR THE WORKS SCHEDULED

Ultimately, the Janina VI deepened shaft will serve as material and personnel transport shaft as well as ventilation intake shaft.

Up to the depth of 800m, material and personnel transport will be will be carried out with counterbalanced large size cage.

The shaft will be deepened from 523 m to 823 m Levels. The following three final intakes will be made in the deepened section of the shaft:
– ventilation station at 756 m Level,
– draw-tube telescope station at 766 m Level,
– both-sided station with basement for the near shaft equipment at 800 m Level.

Construction of the shaft's pipe in the section being deepened is carried out in very difficult geological conditions due to high levels in soil and underground water and very low rock mass strength values.

In the shaft section being deepened, there are six water-bearing horizons featured by high pressure and high content of magnesium and sulphite ions, XA3 exposure class.

Currently, water flow amounts to ca.120 LPM.

Due to high levels in soil and underground water, it was necessary to perform drainage work with perforated pipes to reduce the impact of high pressure aggressive water on shaft furnishing and lining.

Due to difficult geological conditions, VERPENSIN pre-emptive chemical injection is applied in the injection holes.

144

This method consists of drilling eight injection holes in the shaft bottom and pressing injection medium into them to seal rock mass cracks and pores.

Moreover, due to the ground low compactness parameters, it became necessary to make five bearing footings in the shaft lining on the respective depths, in accordance with the guidelines on bearing footings as follow:
– they were located in the rocks with minimum cracking,
– in non-weak rock mass type, with compression strength $R_C > 10$ (value determined in the tests performed),
– they were located out of carbon layers,
– bearing footings were no more than 1.5 m wide each.

C35/45 class concrete was supplemented with specially selected chemical additives due to high water aggressiveness.

Figure 2. Shaft bank before work commencement

4. PREPARATORY WORK ON THE SURFACE

The entire surface of the site is placed on the land not intended for construction. The ground is composed of shale coal, sand, crushed brick, stone, sandy clay and sandstone gravel.

A partial replacement of soil and making continuous footings of large dimensions were required for founding construction objects on the ground.

Besides, piling was needed for the use of P-V temporary headframe.

Piles 28 m long and of 1000 mm in diameter were made under every headframe footing and additional stabilizing piles were made under the shaft staircase structure of 800 mm in diameter.

Figure 3. Area piling

For hoisting equipment exposed to significant congestion and vibrations, there were made concrete slabs for intermediate foundation of the hoisting equipment systems.

The entire construction site was fenced.

Passageways were made of slabs and sidewalks with cobblestones.

The following infrastructure facilities are located within the construction site:
– P-V type headframe with a staircase and discharging chute,
– MPP-9 type hoisting machines (N. and S. points),
– KUBA-10 low speed winches of steel formwork,
– EWP-35 low speed winches of sinking stage-tensioning frame,
– EPR-ESW5T cable winch,
– 6 kV switchgear cabinet,
– cable tray racks,
– 2×10 t telphers,
– workshop and storage building,
– compressor building with compressed air tank,
– container reception,
– car wheels wash,
– social and office container,
– water settling tank.

Figure 4. Headframe installation

There was made a comprehensive installation for machinery and technological equipment power-er supply, for the purposes of the work related to deepening of the shaft.

Construction site is supplied from the mine switchgear to 1000 kVA container transformer sta-tion, made specifically for this project.

This self-propelled transformer station is placed on the foundation blocks. Its primary and backup power voltage supplies amount to 6 kV, 500 V, 400 V and 230 V.

Power cables are routed on cable tray racks.

Figure 5. Construction site after work completion

Figure 6. Shaft bank

5. PREPARATORY WORK IN THE SHAFT

There were also made operating stages at specific levels for the purpose of shaft deepening.

Explosives transport stage with electrically- opened hatches to N. point, was built at 350 m Level.

Creeper is used for materials transportation from depot at 300m Level to the shaft at 350 m Level, where they are loaded into a bucket and lowered to the shaft bottom.

Stage with manually – opened hatches to S. point as well as tanks and dewatering pumps were installed at 500 m Level.

Furthermore, access stage for the dead ends of guide-and-carrying ropes and a three-deck over-flow stage were installed at 650 m Level, in the installation place of tanks and pumps.

This stage is scheduled for liquidation when permanent lining is completed.

TEMIX system adapted to the needs of shaft sinking was used for continuous tension monitoring of guide-and-carrying ropes. The system is based on the solutions used for multi-rope cages.

Sinking stage-tensioning frame is suspended on four ropes, which through rope sheaves mounted on the stage are placed on the girders of rope dead ends at 470 m Level.

Suspension cleats developed by KOPEX – PBSz are equipped with WPS-2/4-20 force measuring inserts, two for each of suspension dead ends.

Tension measurement results of guide-and-carrying ropes are sent periodically to controller where they are visualized on the monitor screen of the winch operator stand.

Thus, the operator obtains a complete data on tension parameters of specific stage ropes.

The operator, thanks to ammeters and visualization of rope tensions, is able to capture irregularities during the platform moving operation and its spreading after setting in the desired position.

In the case of uneven strength distribution, proper rope tension adjustment can be made easily.

Shaft hoist includes the equipment as follows:

– MPP-9 hoist	2 pcs.
– EWP-35 slow speed winch	4 pcs.
– P-V shaft tower	
– PBSz-4 shaft signalling system	
– sinking stage with CŁS loader, as per design	
– DEZAM hatch opening device	
– muck bucket 2.5 m³	2 pcs.
– muck bucket 2.0 m³	2 pcs.
– concrete bucket	2 pcs.

During shaft sinking, there was made a test run of a new model with conical suspension, developed by Kopex – PBSz.

It was carried out in the Janina VI shaft of TAURON MINING Janina Coal Mine in Libiąż in April 2014.

A conical suspension on the rope end of the northern machine of the bucket shaft hoist was flooded with Wirelock resin. The rope was 25 mm in diameter.

During a six-month test run, there was no problem with conical suspension.

Throughout the whole test run, the suspension was subjected to detailed daily audits and examinations by technical-power engineer from KOPEX- PBSz and expert from CBiDGP (Research and Supervisory Centre of Underground Mining Co. Ltd).

After completion of the test run, the suspension was cut off and sent to the Destructive and Material Testing Laboratory in Lędziny, where it passed the tests and can be permitted to the use.

To execute work in the shaft, there was designed and made a sinking stage- tensioning frame.

It is suspended on four EWP-35 low-speed winches and guide hoisting ropes of Ø40 mm in diameter are wounded on them.

It is used for:
– shaft sinking,
– making shaft lining,
– installation of pipelines,
– emergency works.

A sinking stage -tensioning frame with CŁS loading system was designed as a three-platform version. Climbing between the platforms is ensured by a ladder with safety cages. The distance between every platform is 4.5 m.

Platforms are made of rolled sections joined by six bearing pillars with fish-plates and pins.

Under the top platform, there are installed rope sheaves Ø900 mm in diameter, through which guide-and-carrying ropes from winches to binding joists are led. Rope dead ends are fixed there.

Between the top and bottom platforms there is made a circular cross-section ventilation system for guiding air-ducts, lowered below the sinking stage-tensioning frame.

– *Top platform* – designed in a diameter of 6,600 mm.

Hatches opened with hydraulic cylinders were installed in the bucket passages. Passages were designed for bucket volumes V = 2.5 m^3 or 2.0 m^3.

Along two edges of the hatches there were fastened protective barriers matching concrete hinged chute.

Besides, guard rails 1,200 mm high were also fitted to the platform edges.

The platform is stabilized by four spragging actuators.

To protect personnel, protective roofs were installed in the top platform. In addition, four lamp stands have to be installed on the platform, too.

– *Central platform* – designed in a diameter of 6,600 mm.

Enclosures were installed in the bucket passages.

On the platform, there were located water tanks, hydraulic set and transformer distribution unit.

A pump pressing water into the mine drainage network was disposed in one of the tanks.

Guard rails 1,200 mm high were also fitted to the platform edges.

The platform is equipped with electrical installation for lighting platform.

– *Lower platform* – designed in a diameter of 6,230 mm.

Enclosures were installed in the bucket passages.

On the stage perimeter, there is mounted a tilt-flaps seal with a rubber band, matching the shaft housing built.

The platform is stabilized by six spragging actuators.

Bottom part of the platform ring is adapted for fixing the CŁS loader raceway, and loader arm central support was located in the geometric centre of the platform.

There were also designed manholes in the platform to climb down to a hanging ladder and into the CŁS loader cabin.

In addition, the platform was equipped with hydraulic equipment and installation as well as with electrical installation for lighting the platform and the shaft bottom.

Sinking stage-tensioning frame is equipped with a hydraulic system used for spacing platform in the shaft, hatches drive on the top platform as well as for spacing steel formwork, concrete separation device and the lighting and signalling installations.

Technological equipment on the sinking stage-tensioning frame is supplied with power cable mounted in the shaft gradually with work progress using shaft cable cleats suspended at intervals not longer than 6 m.

Cable reserve is collected on the central platform of the sinking stage-tensioning frame.

Besides, fire protection equipment was located there in accordance with Item 5, Attachment 5 to the Decree of the Ministry of Economy dated 28 June 2002 (Journal of Laws No. 139, Item. 1169), due to the use of oil in the hydraulic system.

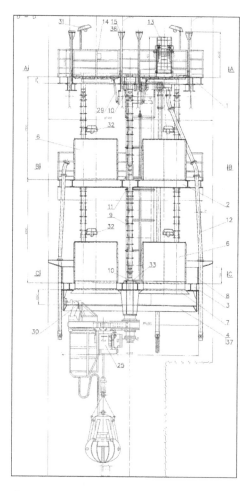

Figure 7. Sinking stage – tensioning frame

6. COMMUNICATION AND SHAFT SIGNALING

The SLS intrinsically safe shaft communication system mounted in the PBSz-4 signalling and shaft communication equipment is used for communication.

It enables communication among all signalling stations for supporting the GWSz.

Signalling stations are located in the shaft bottom, sinking stage, at 500 m and 350 m Levels, shaft bank as well as on discharge stages and rope sheaves.

One strike direct signalling is used in communication between shaft signalman and hoist operator.

Execution signals are broadcast using execution transmitters built-in in every signal station.

A button on the ECHO-FG keypad device is used as transmitter in a bucket.

Execution signals from the sinking stage are broadcast using a button on the cage camera of the ECHO-PG device.

Execution and alarm transmitters installed on the sinking stage, broadcast signals from the shaft bottom.

Signal cords are attached to transmitters levers and lowered to the signal station at the shaft bottom.

From this station, signals are broadcast sent to the steel formwork and a bucket between the deck and the shaft bottom.

During shaft accessing from the bucket or personnel transportation, communication with hoist operator is possible using the ECHO-FG device.

In addition, it is possible to communicate with wireless radio phones approved for use in underground mining conditions.

In case of failure of any of the devices: ECHO-PG or ECHO-FG, what is indicated by an alarm signal, broadcasting signals from a specific position on the sinking stage or in the bucket is possible with signal cord connected to transmitter lever on the shaft bank.

Signal- and- alarm cord allows for broadcasting signals from the bucket by suspended crossheads or the ECHO-FG device failure and it is used during work execution in the shaft.

Indicator lights are used for optical signalling from hoist operator's place positions of shaft hatches at 350 m and 500 m Levels.

Signalling device also allows for mating the device located in the second shaft compartment.

Monitoring system based on external cameras with IR lighting was installed in the hoist cabins.

Cameras allow for monitoring the shaft bank, discharge stage and concrete distribution stage in the shaft.

Monitoring system applied enables hoist operators to observe the GWSz hotspots in order to avoid dangerous situations.

7. VENTILATION IN THE SHAFT SECTION BEING DEEPENED

Air supplied by shaft ventilates 350 m and 500 m Levels of the mine.

Shaft forehead is ventilated by suction ventilation.

Flexible shaft suction vent pipes of 1000 mm in diameter and flexible supply air-pipes to intensify ventilation after blasting were used in the mine.

Air-pipes 10 m long are joined with each other by screw joint and are suspended on the brackets in the shaft.

Stale air is sucked out of the shaft forehead and directed into stale air- duct in the dip road at 500 m Level.

Length of the air-duct is gradually extended as long as work is progressing, to obtain the right amount of air for maintaining suitable climatic conditions underground.

A distance of minimum 16m between air-duct and the shaft bottom is kept.

To ensure the right amount of air exchanged at a maximum length of the air duct, the WLE 1005 B/CZ fan was applied.

Whereas, the 813B/E/1 air-duct fan, being an auxiliary device for shortening ventilation time, was installed in the upper part of the sinking stage- tensioning frame, next to the built-in descending ladder.

It is used for providing additional ventilation, e.g. after blasting. Supply fan capacity is 6.7 m^3/s and is by 33% less than the suction fan capacity (10 m^3/s).

When a very large break-out is made, additional ventilation is also needed. The amount of air being supplied is continuously monitored by air flow sensor installed at 515 m Level of the shaft.

The following rules are obeyed at work execution:
- end of the duct line should always be below the sinking stage (except for the installation – duct line extension),
- work should be carried out so that the end of the duct line is not higher than the maximum calculated value, i.e. 16 m from the shaft bottom.

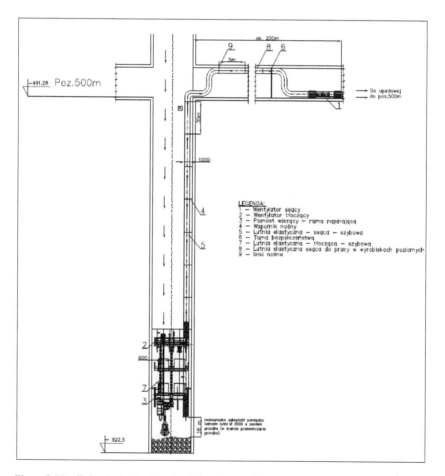

Figure 8. Ventilation in the shaft section being deepened

In winter, shaft heating is carried out with 4 heating panels.

This allows for heating shaft to the temperature of +2°C at minimum external air temperature of –20°C.

Power heaters are provided with heating medium from the mine heating network.

Heating panels are placed parallelly on foundation plates, in the north-eastern part of the shaft. They are connected to the collector with steel passages of dimensions 1,000×1,000 mm.

The air supplying panels is pulled from the atmosphere through chimneys. Its quantity is regulated by blinds installed in the panels and chimneys.

8. DEEPENING THE SHAFT

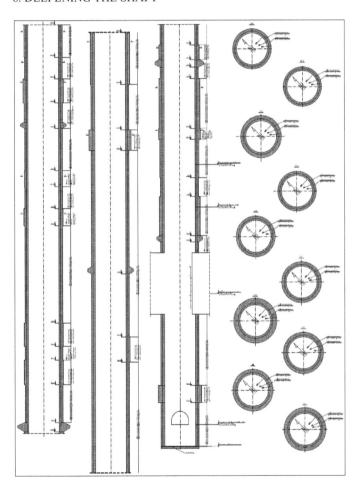

Figure 9. Lining sections in the shaft being deepened

Deepening the Janina VI shaft in temporary lining (external column of combined lining) is carried out to the technological station of the ventilation gallery designed at the depth of 780 m, and subsequently, a permanent lining with slipper from the bottom to the top will be constructed.

153

Combined lining at the depth of 291.1 m of the shaft being deepened is made of C35/45 class concrete, where:
– outer column was made as concrete lining of thickness ranging between 0.45 m and 0.90 m,
– inner column was made with concrete, 0.45 m thick.

Combined lining made in C40/50 class concrete with outer column 1.10 m thick and inner column 0.45 m thick were designed for the remaining part of the shaft, at the depths between 641.95 m and 646.85 m as well as between 808.15 m and 812.15 m.

Lining thicknesses are sufficient for anchors for fixing shaft furnishing and DN100 water run-off pipeline in internal column concrete. A pipeline, ca 250 m long was installed in the temporary lining.

Considering difficult geological conditions in the entire shaft's tube, combined lining in the shaft section scheduled for deepening was designed in a specific way.

The outer column can work independently during shaft deepening and "n" – its certainty factor of stress transfer increased to the value of 0.8 for the 2nd degree hazard for shaft lining, ranging between 1 and 0.75 – will be maintained.

If during shaft deepening some varying geological conditions or deposition depths of specific layers, which may have a significant impact on the calculated load shaft lining are stated, there will be made corrections of calculations and shaft lining selection.

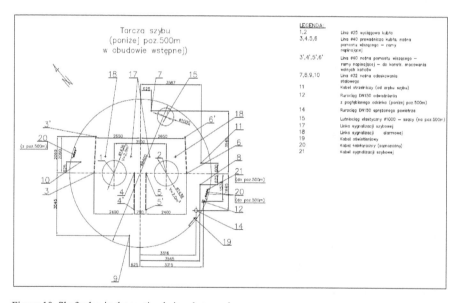

Figure 10. Shaft plan in the section being deepened

Deepening the shaft is an operation carried out in a cyclic manner.
Shaft deepening includes the following stages:
– muck removal and shaft bottom cleaning,
– drilling blast holes in accordance with blasting card,
– blasting,
– muck extraction,
– setting steel formwork,
– making concrete shaft lining.

Figure 11. Shaft bottom view before deepening commencement

In the first phase of shaft deepening, in preparatory work, there was made reinforcement with 48 pieces of POK-22 anchors arranged alternately in two rows, over the scheduled breaking-out shaft lining, to strengthen and stabilize rock mass.

As it was impossible to suspend a sinking stage- tensioning frame in this phase of work, extracting muck was executed with GRYF-1P shaft loader, mounted on KCH-9 winch.

'The GRYF-1P is operated by the operator from the shaft bottom. A lot of experience in work related to shaft sinking is required from him. Muck is extracted on the discharge chute by buckets of 2.5 m³ capacity.

Emptying the bucket is done using bucket discharge system.

In this phase of shaft sinking, blasting was carried out at the depth of 2 m, due to a close distance to the structures built-in in the shaft as well as at the depth of 500 m.

For safety reasons, steel formwork is lowered systematically as muck extraction progresses, to prevent rocks sliding from the shaft cheek that could cause rock falling on operating miners.

When muck excavation at the height of steel formwork is completed and the dimensions of breakout are checked, steel formwork is lowered to the vertical centre and spragged.

Pouring concrete was done with concreting hopper mounted in the stage passage at 500 m Level.

Concrete was delivered with buckets to 500m Level where it was poured with a hose attached to the hopper into ingots in steel formwork.

Buckets with bottom opening were attached to suspension gears in the shaft bank.

The buckets are filled with concrete on the concreting stage, at –12 m Level where servicemen pour concrete into buckets placed in the proper position.

The PSz-125 pump was used for shaft dewatering in the first phase of its sinking. Water was pumped with a hose of 100 mm in diameter into mine drainage system at 500 m Level.

Whereas at muck extracting and blasting, water was pumped by Depa DL-80 pump to a bucket and directed to the surface.

After installation of the sinking stage-tensioning frame, sinking the shaft was commenced with the use of the CŁS heavy duty loader, equipped with high-capacity buckets what significantly speeded up extracting muck.

As during blasting, the sinking stage-tensioning frame had to be departed from the shaft bottom in a safe distance of minimum 20 m, it was installed at the depth of ca. 540 m.

Blast hole drilling is carried out at the depth of 4 m in accordance with blasting card.

Shaft technological furnishing is successively being made in work progress.

Technological furnishing includes pipelines as follows: compressed air Ø150, drainage Ø150, fire safety Ø100, suction air duct Ø1000 and power cable to the sinking stage, extended in 60 m pieces.

Signalling and telecommunication cables were collected in the middle floor of the sinking stage due to their relatively low weight.

In this phase of shaft sinking, concreting is carried out with concreting buckets with the use of concrete separation equipment built-in on the upper platform of the sinking stage-tensioning frame.

After opening the bucket bottom, pouring concrete into the behind the steel formwork starts.

When deepening the shaft below the ventilation station, installed at the depth of ca 780 m is completed, steel formwork of 7.5 m in diameter will be made and inner concrete lining from the bottom to the top.

The shaft will be further deepened in temporary lining, up to the final depth (in the same way as in the section above), and permanent lining will be constructed from the bottom to the top.

For this purpose, first a steel framework will be mounted on the final shaft diameter of dimension 7.5 m.

Permanent formwork has to be set in preliminary lining directly above the preliminary formwork left in the shaft. Permanent formwork will be sealed from the bottom and the work will be executed from the operating stage, made in preliminary formwork.

When the required sprigging and setting formwork are made, the first from the bottom section of the permanent lining will be concreted.

Concreting is made in the same way as in the case of preliminary lining. Subsequently, a stage for installing formwork will be assembled.

It will be followed by constructing continuously the permanent lining.

This mode of constructing the permanent lining is called "slipper".

Mining Forum 2015, Kicki et. al. (eds) © 2015 Taylor & Francis Group, London, UK. ISBN 978-1-138-02820-3

Vertical Position Identification of the Machine Vision System in the Mineshaft for the Purposes of Diagnostic Image Processing of the Mineshaft Lining Surface

Henryk Kleta, Adam Heyduk, Jarosław Joostberens
Technical University of Silesia, Gliwice, Poland

ABSTRACT: The paper presents some problems of mineshaft technical condition assessment. These problems are particularly important in the case of mineshaft with no hoisting machines. There has been described a detailed example of insulating dam technical condition analysis. Particular attention has been paid to scale coefficient values for conversion from image dimensions to real world dimensions. Because values of these coefficients heavily depend on the distance between camera and the lining surface, a vertical position identification algorithm has been updated with a distance compensation formula.

KEYWORDS: Mineshaft inspection, shaft monitoring, isolating dam assessment, position identification

1. INTRODUCTION

Mineshaft technical condition analysis is based on a macroscopic evaluation of its lining, therefore it heavily depends on the lining accessibility. Where there is no hoisting machine in the shaft macroscopic evaluation is significantly reduced, due to the lack of full access to the shaft lining. In these cases the shaft lining technical condition analysis is reduced only to lining macroscopic assessment, performed with mobile rescue lift or is based on a visual lining assessment on the basis of the video signal – recorded by a camera system lowered into the shaft (Fig. 1).

In the case of large shaft lining damages, it is impossible to carry out direct checks on the damaged sections of linings, even with the lift rescue. In these circumstances, it is necessary to use monitoring solutions based on the use of the measuring apparatus, moving in the shaft without the presence of humans. In this case, for a reliable assessment of the shaft lining technical condition, it is very useful to use modern computer image analysis software together with numerical computations for identifying the stress status and lining damage extent [1].

Computer-assisted visual analysis is one of the most convenient methods, particularly in the terms of human workload cost and measurement time demands. It is used to evaluate the technical condition of the concrete objects such as railway tunnels, road surfaces, and bridges. Machine vision methods of vision have been also applied to detect defects in the interior of underground pipelines.

2. AN EXAMPLE OF ISOLATING DAM TECHNICAL CONDITION VISUAL ANALYSIS

Mineshaft X has been adapted to perform the function a submersible pump station. It has been partially sunk, and the water level is maintained in the depth range of 423.5÷393.5 m.

In the mineshaft X, with no hoisting machine – at the level of 327 meters there has been is located a shaft station closed with insulating brick dams situated in the inlet to the shaft tube (Fig. 2). Excavation area behind the dams is full of water and through leaks in dams and rock cracks around them, three overflows water, which flows into the shaft. Completely watered insulation dam Ti-2 at the 327 meters level under the continuous high water pressure, is a threat to the built-in pump routes of shaft submersible pump station arranged at a depth of approx. 440 m.

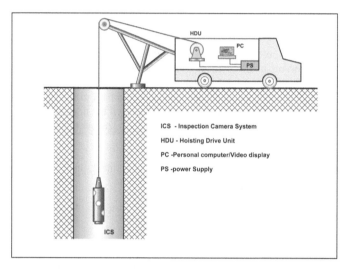

Figure 1. Schematic of a mobile video inspection system
for shafts without hoisting machines [1]

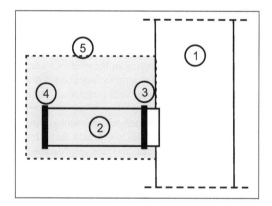

Figure 2. Condition diagram of the X shaft at 327 level: 1 – shaft , 2 – shaft station
at 327 m level, 3 –Ti-1 dam, 4 –Ti-2 dam, 5 – level of accumulated water

158

In the case of the breakage of the dam described above – by the water pressure – it is possible a sudden intrusion of an amount of water or water with a loose material, as well as fragments of broken Ti-1 and Ti-2 dams into the shaft tube. It can lead to difficult to assess damages and pump station stoppage.

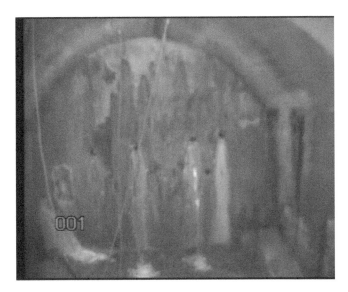

Figure 3. An insulating dam view –a selected video frame from the June 2013

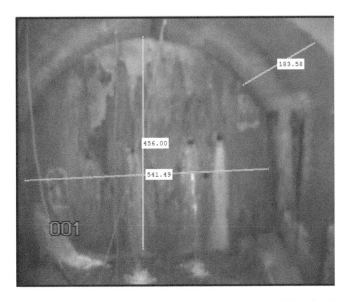

Figure 4. Characteristic distances of holes in isolating dam at 327 m level measured (in pixels) directly on the camera recorded frame – a basis for further calculation of scaling coefficients

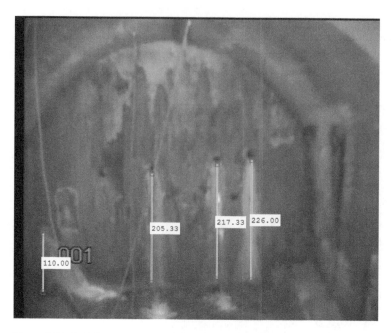

Figure 5. Vertical distances of holes in isolating dam at 327 m level
measured (in pixels) directly on the camera recorded frame

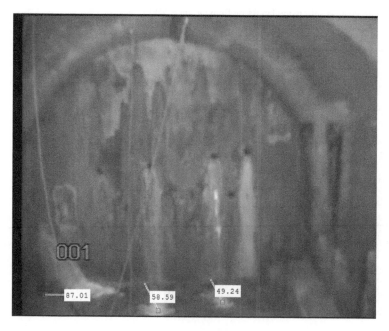

Figure 6. Horizontal distances of holes in isolating dam at 327 m level
measured (in pixels) directly on the camera recorded frame

All the distances are measured on image frames in pixel units (because pixels are the smallest image element), but for the further analysis they have to be converted into real world length units (m, cm, mm etc.). It can be achieved on the basis of triangle similarity and Thales theorem (Fig. 7), using known global working dimensions (height, width, length). These conversion coefficients have been presented in Table 1.

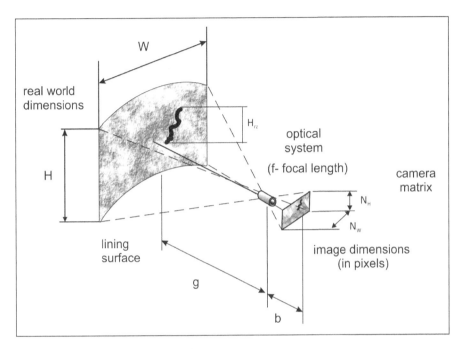

Figure 7. Triangle similarity and scaling between
real world dimensions and image dimensions

Table 1. Scaling coefficients for image dimensions conversion

	Image dimension	Real dimension	Scaling coefficient
height	456 pix	3,1 m	6,8 mm/pix
width	541.5 pix	2,7 m	5 mm/pix
depth	183,5 pix	2 m	10,9 mm/pix

After considering the scaling coefficient values (calculated separately in three orthogonal dimensions for the entire excavation cross section) there can be estimated hole diameters and positions and range of the water stream flowing from behind the insulating dam at 327 m level. This range is tightly connected with the water pressure.

Thus determined the geometrical dimensions and location of holes in the insulating dam on the 327 m level allow us to assess the height of water accumulation behind the dam, and the successive analysis allows for the validation of possible height changes in the accumulated water amount behind the insulating dam at 327 level.

2. IDENTIFICATION OF VERTICAL CAMERA POSITION
   BASED ON VIDEO STREAM ANALYSIS

In the case of mineshaft lining video monitoring systems, there is an important issue of accurate identification of the actual position of the currently displayed or processed part of the video stream (frame). This is due to the fact that the for the visual lining state assessment it is needed not only an information about the possible shaft lining damage (e.g. crack), but also an information on its position inside the shaft lining surface – especially when comparing images made in a large interval. Due to the relatively short (in comparison to the camera movement speed) interval between the frames and therefore an associated high cross coverage rate for neighboring frames, it is possible to calculate a spatial offset of two adjacent frames using a cross-correlation method. The components of this offset are proportional to the camera instantaneous speed, so summing up all these elementary shifts (corresponding to integrating the instantaneous velocity function) allows us to estimate the current camera position in the mineshaft. Detailed considerations on determining the mutual shift of two adjacent video frames using the two-dimensional cross-correlation function have been shown in [2].

Mutual shift of two images is determined in pixels, which are the basic spatial elements of a digital image. An actual length of mapped lining fragments $\Delta X, \Delta Y$ can be determined on the basis of displacements expressed in pixels $\Delta x, \Delta y$ and the knowledge of the scaling factor C [mm/pix] as:

$$\begin{cases} \Delta X = C \cdot \Delta x \\ \Delta Y = C \cdot \Delta y . \end{cases}$$

The value of C coefficient can be determined with knowledge of the camera optical parameters of (focal length f), the camera sensor (Nx, Ny pixels, diagonal D) and the distance of the camera from the base of the observed surface b as:

$$C = \frac{D}{\sqrt{N_x^2 + N_y^2}} \cdot \left(\frac{b}{f} - 1\right).$$

Base distance b can be determined using theoretical assumption of a centric camera position in the mineshaft (half the diameter of the shaft reduced by half the diameter of the camera capsule). It should be noted, however, that in practice, the task can be difficult due to the non-centric position of the capsule in the shaft and the associated unequal optical distance of each camera from the mineshaft center (Fig. 9).

In the view of the above-mentioned problems, an elementary movements aggregation process may cause quite significant errors in determining the position of the camera capsule. In [2] there has been analyzed an impact of changes in the camera's distance from the lining surface on the accuracy of the recorded video playback and the camera position estimation. Using distance sensors (eg. a laser or ultrasonic) is however possible to continuously adjust the scaling factor values in the algorithm of the camera position identification.

The Thales theorem and the triangles similarity theorem (Fig. 9) show that for a given – different from the base one – distance of the camera from the lining surface on the adjusted (real) value of C(r) can be determined based on the value of the base value C_b as:

$$C(r) = C_b \cdot \frac{r}{b} .$$

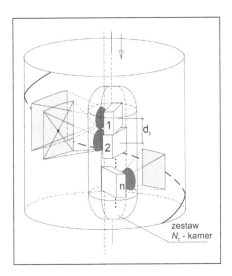

Figure 8. Non-centric location of the camera capsule
in the mineshaft cross section area

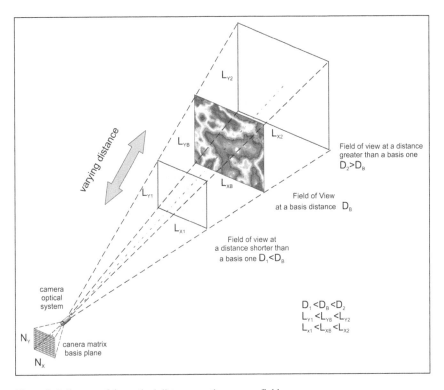

Figure 9. Influence of the optical distance on the camera field
of view and scaling coefficients

Vertical capsule position identification algorithm complemented by distance compensation has been shown in Figure 10.

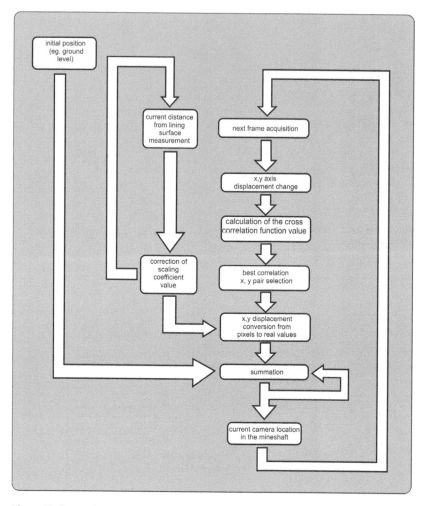

Figure 10. Camera location identification algorithm with variable distance from lining surface

CONCLUSIONS

Machine vision assisted analysis can be very useful in the lining technical condition assessment particularly in the case of mineshafts without hoisting machines (e.g. mineshafts converted into deep wells). The scaling coefficients are necessary for the conversion from image dimension measurements (in pixels) to real world dimensions (in meters). But their values heavily depend on the distance between camera and the surface lining. Therefore for the precise identification of vertical camera position – using a cross-correlation method – it is desirable to use this distance value in a compensation formula.

REFERENCES

[1] Kleta H., Heyduk A.: Image Processing and Analysis as a Diagnostic Tool for an Underground Infrastructure Technical Condition Monitoring in Underground Infrastructure of Urban Areas 3. (Madryas C. and Szot A. eds), pp. 39–46, A.A. Balkema, Rotterdam 2014.
[2] Kleta H. (ed.), Bączek A., Chudek M., Cierpisz S., Heyduk A., Jendryś M., Joostberens J.: Visual Method of Mine Shaft Lining Technical State and Safety Assessment Using Digital Image Analysis. Wydawnictwo Politechniki Śląskiej, Gliwice 2013.

International Mining Forum 2015, Kicki et. al. (eds) © 2015 Taylor & Francis Group, London, UK. ISBN 978-1-138-02820-3

Sinking of 1 Bzie Shaft in Difficult Hydrogeological Conditions with the Application of Combined Lining and High-Performance Concretes

Daniel Wowra
Kopex – Przedsiębiorstwo Budowy Szybów S.A. Bytom
daniel.wowra@kopex.com.pl

Tomasz Sanocki
Kopex – Przedsiębiorstwo Budowy Szybów S.A. Bytom
tomasz.sanocki@kopex.com.pl

Mariusz Wojtaczka
Kopex – Przedsiębiorstwo Budowy Szybów S.A. Bytom
mariusz.wojtaczka@kopex.com.pl

SUMMARY: Since 2009 Kopex – Przedsiębiorstwo Budowy Szybów S.A. (Kopex – Shaft Sinking Company) has been carrying out sinking of 1 Bzie shaft for Borynia-Zofiówka-Jastrzębie Coal Mine belonging to Jastrzębska Spółka Węglowa S.A.

The paper presents meaning of the proper selection of sinking method and type of shaft lining in difficult geological conditions of the deposit overburden. In relation to the occurrence of formations of low-strength parameters, especially in the conditions of watered and gasified rock mass, the special methods were applied: fore injection for consolidation and strengthening, ground rods, drainage holes. Application of proper methods and technological disciplines enabled the safe shaft sinking at the most difficult section in the deposit overburden.

Accordingly to the difficult hydrogeological conditions, an appropriate structure of combined shaft lining was applied: panel-concrete and steel-concrete using the high-performance concretes.

KEYWORDS: Difficult hydrogeological conditions, shaft sinking, special methods, shaft lining

1. INTRODUCTION

Detailed examination of the rock mass watering and gasification, the magnitude of formation pressures, the quantity of water and gas inflows to the excavation as well as the engineering-geological parameters, followed by proper classification of the geological conditions provide the basis for selecting a suitable special method for the preliminary modification of unfavourable water/gas and strength properties of the rock mass.

When exploited in a weak, watered rock mass, mine shafts forming the main opening-out excavations require – for secure functioning – in particular the application of a proper shaft lining that is adequate to the actual geological conditions. Proper designing of the shaft lining and the application of a suitable sinking technology are difficult and unrepeatable, because the great number of strictly-related geological, technical and technological conditions requires an individual approach to each particular excavation through the analysis of all the factors and the dependencies between them.

During the sinking of the 1 Bzie shaft in the Borynia-Zofiówka-Jastrzębie Coal Mine belonging to Jastrzębska Spółka Węglowa S.A. in weak overburden formations of very great thickness, which were simultaneously watered and gasified, there were conditions requiring the application of a combination of several special methods for the preliminary modification of the rock mass properties: fore consolidation and strengthening of the rock mass by injection agents, stabilisation by ground rods, as well as drainage of water and gas.

Based on the experience gained, the article presents the methods applied in order to de-stress and strengthen the weak rock mass. The article also describes the solutions applied for the construction of the preliminary shaft lining made of precast reinforced concrete and steel segments as well as of the final lining made of high-performance concretes under high technological rigours.

2. DIFFICULT GEOLOGICAL CONDITIONS DURING THE SINKING OF THE 1 BZIE SHAFT

In the analysed region of the 1 Bzie shaft, where the sinking is carried out in the south part of the Upper Silesian Coal Basin, the difficult geological conditions are closely related to the high thickness of the deposit overburden formations characterised by low strength parameters, we well as sectional watering and gasification.

In this region, the carboniferous deposit formations lie under the overburden of quaternary formations with thickness of approx. 25 m, as well as cohesive and loosely-compacted tertiary Miocene formations with thickness of approx. 720 m, which contain sixteen water-bearing and gas levels. The tertiary (Skawina beds) forms a very monotonous complex of clays that – along with the increasing depth – turn into claystones with thin interbeddings and laminae of silts, silty sands, fine sands and loosely-compacted sandstones.

Traces of the Dębowiec layer formations were found in the floor part of the tertiary, which have been untypically developed as conglomerates of sandstones and claystones, whilst the carboniferous roof has been developed in the form of claystones and sandstones with interbeddings of mudstones, coal deposits, and inserts of tectonic breccia. A zone of concentrated tectonic disturbances was found in the floor part of the tertiary formations from the depth of approximately 710 m and in the carboniferous roof.

The sinking shaft cut through three water-bearing levels in quaternary, tertiary and carboniferous formations.

Due to the thickness of the complex and the total volume of the inflow, the key roles is played by the tertiary level, in which watering was exhibited by the dense, thin interbeddings of silts, sands and loosely-compacted sandstones with low filtration coefficient as well as by the approx. 2÷3 meter layer of a loosely-compacted sandstone stratified with sand (benchmark sandstone). The maximum pressures in the water-gas horizons were 7.4 MPa, while the measured maximum methane inflows in the exploratory well were above 20 m³/min.

In the thick mass of tertiary rocks, the sinking was generally carried out in non-rock formations not resistant to water, and therefore their strength parameters decreased with the increase of humidity. It was only in the lower part of the Tertiary that the cohesive clays of generally compacted and segmentally plastic consistency began to turn into loosely-compacted claystones. The clays and claystones contained thin interbeddings and dense lamination of loose sands, cohesive silts and sandstones from loosely compacted to compacted ones, which further weakened the rock mass.

The carboniferous roof was dominated by highly-eroded clay formations and therefore their water resistance was also low. Additionally, over the sections affected by tectonic disturbances, the rock mass was heavily and very heavily cracked, which caused a high risk of rocks falling off the sidewall.

A significant improvement of the sinking conditions occurred in the section of the residual carboniferous rocky formations, particularly in the compacted sandstones below the depth of approx. 765 m, which are characterised by a considerably increased resistance to water.

The formations occurring in the analysed area of sinking the 1 Bzie shaft at the section of the deposit overburden are generally not resistant to water. Sectionally, the rock mass exhibits dense stratification, lamination and joints. Over the sections affected by tectonic disturbances, the rock mass was heavily and very heavily cracked.

These factors, and the lack of water resistance in particular, led to gradual reduction of strength parameters and the loosening of the sidewall rocks.

As a result, the rings of the preliminary ground support were closed as quickly as possible, in order to minimise the stress-relieving cracking and water migration, and thus to prevent the rapid degradation of the rock mass. Also the space between the preliminary lining and the rock mass was filled with high precision to ensure good cooperation and limit the water flow.

Tertiary (Period)				Quaternary (Period)	1	STRATIGRAPHY
V ~50 m	IV ~80 m	III ~110 m	II ~210 m	I ~25 m	2	ROCK MASS ZONES
Loam//loosely-compacted sandstone, silt and silty sand			Loam//sand and silt	G//sand	3	LITHOLOGY
			single water infiltrations		4	WATER-BEARING LAYERS
9×10^{-8}	$2\text{-}7 \times 10^{-7}$	no exploratory pumping was performed			5	FILTRATION COEFFICIENT [m/s]
~4,6-4,7	~1,6-2,8	~2,7	-	water table	6	WATER PRESSURE [MPa]
up to several m³/min	small inflow	very small inflow	none	none	7	METHANE
CO found	none	none	none	none	8	CO I CO_2
0-57 (38)	0-49 (17)	core not collected			9	(RQD) ROCK QUALITY DESIGNATION
< 1,5 h	< 1,5 h	< 4 h	< 24 h	-	10	SOAKING TEST OF ROCKS
inflow to the shaft with sinking < 290 l/min together with the drainage hole	inflow to the shaft with sinking < 268 l/min together with the drainage hole	inflow to the shaft with sinking < 130 l/min	from ~80 m minor water infiltrations during shaft sinking	-	11	NOTES

geological conditions on the basis of exploratory hole BD 57BIS

Carboniferous (Period)	Tertiary (Period)					STRA-TIGRA-PHY
X ~85 m	IX ~85 m	VIII ~3 m	VII ~100 m	VI ~80 m	2	ROCK MASS ZONES
Sandstone// //mudstone and claystone	Claystone//sandstone loosely compacted	loosely-compacted sandstone	Loam-Claystone// //loosely-compacted sandstone	Loam//loosely-compacted sandstone and silty sand	3	LITHO-LOGY
					4	WATER-BEARING LAYERS
3×10^{-9}	$2\text{-}4\times10^{-8}$	7×10^{-7}	1×10^{-7} 4×10^{-8}	$1\text{-}9\times10^{-8}$	5	FILTRA-TION COEFFI-CIENT [m/s]
~6,3	~7,0-7,4	~6,4	~5,8-6,5	~5,3-5,6	6	WATER PRES-SURE [MPa]
small inflow	small inflow	small inflow	small inflow	up to a dozen m^3/min	7	METHA-NE
CO found	high CO concentration	CO found	CO found	CO found	8	CO I CO_2
0-70 (30)	20-91 (46)	-	10-76 (52)	19-81 (50)	9	(RQD) ROCK QUALITY DESIGNA-TION
AAAA by Skutty	< 6 hours	< 15 minutes	< 1,5 hours	< 2,5 hours	10	SOAKING TEST OF ROCKS
inflow to the shaft with sinking < 242 l/min	inflow to the shaft with sinking < 255 l/min	inflow to the shaft with sinking < 260 l/min	inflow to the shaft with sinking < 595 l/min together with the drainage hole	inflow to the shaft with sinking < 295 l/min together with the drainage hole	11	NOTES

geological conditions on the basis of exploratory hole BD 57BIS

Figure 1. Scheme of geological conditions on the basis of exploratory hole for 1 Bzie shaft sinking

3. SPECIAL METHODS OF SHAFT SINKING APPLIED.

The application of a combination of several special methods during the sinking of the 1 Bzie shaft was closely related to the presence of difficult geological conditions. Because it was necessary to ensure safety of the shaft sinking works performed, protection against falling sidewall rocks and proper construction of the shaft lining, preliminary technological solutions were applied in order to de-stress the rock mass, and then consolidate it and stabilise the area around the shaft forehead.

The main special method applied was the fore drainage of water and gas performed through the shaft lining. It was intended to reduce the pressure exerted by the rock mass, by controlled drainage of water and gas from this rock mass. The complementary special methods applied in order to preliminarily modify the properties of the rock mass consisted in the fore injection of quick-setting agents and the stabilisation of the sidewalls by means of ground rods. The technology of ground rod driving and fore injection was based on the drilling method from the bottom of the shaft through the previously prepared concrete screed. The procedures improving the strength parameters of the rock mass in the area of the shaft's tube were performed to the maximum distance of 12 m. The application of the combined special methods in the difficult geological conditions, in the presence of cohesive clay formations with interbeddings and dense laminations of loose formations containing watered and gasified sections, made it possible to construct the 1 Bzie shaft by safely minimising the risk of falling sidewall rocks, and consequently preventing the damage of the shaft lining.

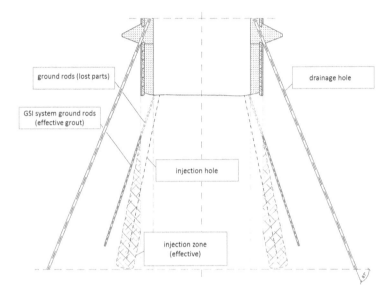

Figure 2. Special methods – drainage and the rock mass strengthening

3.1. *Drainage of water and gas by means of protection holes*

The application of the fore drainage of the rock mass through the drilling wells made in the bottom of the shaft resulted in the formation of a stress-relieving barrier around the sinking shaft due to the controlled drainage of water and gas.

The number of openings drilled was 1÷8, depending on the intensity of watering and gasification of the rock mass. Additionally, geological observations were carried out in the drainage holes to further detail the geological data obtained from the exploratory well drilled from the BD-57bis surface in order to specify the current status of watering and gasification of the drilled layers in the close proximity to the shaft's tube.

The openings were located symmetrically along the perimeter of the shaft. The drilling was performed through casing pipes of 15÷30 m in length, at an approximate angle of -67°. The length of the openings was up to 120 m, and the diameter of the working section of the opening amounted to approximately 65÷150 mm. Due to the water and gas hazards after cementing the casing pipes, a tightness test was performed, and further sections were drilled through the protective high-pressure fittings (gland and bolt) installed at the outlet of the opening.

In order to minimise the gas hazards for the sinking 1 Bzie shaft, advance methane drainage of the rock mass was performed at the depth from 425 to 616 m. The methane drainage was performed using an opening drilled from the BD-57bis surface, a pipeline and a methane drainage station.

3.2. *Consolidation and strengthening of the rock mass using fore injection*

The application of fore injection to the rock mass through the drilling wells made in the bottom of the shaft resulted in the binding of weak formations with quick-setting agents, thus ensuring stabilisation in the shaft area. Moreover, the injection reduced the rock mass permeability in the zone directly by the shaft sidewall, where the formation process of stress-relieving cracking, soaking and loosening of the ground was the most intensive. Outside the sealed zone, the drainage was performed through the above-mentioned drilling wells.

Fifty four (54) injection openings of approx. 75 mm in diameter were drilled symmetrically along the perimeter of the shaft.

The length of the openings was up to 12 m, and the drilling angle was approximately -80°. The injection process was conducted by inserting injection pipes with expansion heads to seal the outlet of the opening. The rock mass was stabilised and reinforced using agents with the following characteristics: high water resistance, good waterproofing properties, high adhesion rate, low viscosity, and high strength. During the sinking of the 1 Bzie shaft in the overburden formations, the injection of the rock mass was carried out using two-component polyurethane adhesives, e.g. the Erkadur®/Erkadol® which is based on resin and a hardener, as well as a mineral binding agent with quick-setting additives, e.g. SPEC-INIEKT 90 NC.

3.3. *Rock mass strengthening by means of ground rods*

Ground rods were applied, using the GSI GONAR system, to stabilise the sidewalls during the sinking of the 1 Bzie shaft in the sections with excessive loosening of the sidewall formations that made it impossible to safely perform the works and construct a proper shaft lining. The stabilisation system of the rock mass based on the ground rods consisted of two operating elements: the reinforcement (rods made of high-grade steel) and the grout body. In order to construct a palisade of ground rods, 51 openings with the diameter of $\phi 76 \div 135$ mm were drilled; they were located symmetrically along the shaft's perimeter. The length of the openings was $3 \div 12$ m, and the drilling angle was approximately -70°. R32N or R51N rods were installed in the openings. They were glued by means of an injection agent, e.g. the Erkadur®/Erkadol® polyurethane adhesive, thus forming the grout body. The migrating injection grout provided additional reinforcement for the ground base around the ground rod under construction.

4. APPLICATION OF COMBINED SHAFT LINING

With a view to safely sinking the 1 Bzie shaft in the deposit overburden section characterised by difficult geological conditions and to facilitating the proper construction of the planned shaft lining, a series of combined special methods for shaft sinking was used to reduce the expected water inflow to the shaft shortwall and the pressure from water and gas exerted on the rock mass in the direct shaft area. Despite using the special methods, it was not possible to completely avoid the water flow in the direct shaft area. Consequently, we experienced – to a limited extent – soaking and loosening of the rocks from the sidewall. For the above reasons, the solution applied in the analysed section of the watered and gasified tertiary formations was the preliminary ground support.

It was constructed from rings made of precast segments, which were installed within the shortest possible time in order to close the ring and shield the sidewall. A very large emphasis was also put on precise filling of the space between the preliminary support and the rock mass with concrete.

The sinking of the Bzie 1 shaft was carried out in the section of deposit overburden and in the carboniferous roof, using the top-to-bottom technology by short 2÷2.1 m high sections. In the first place, the preliminary ground support was constructed in the form of 2÷4 rings. The height of the rings was ranging from 0.5 to 1.0 m, and was dependent on the type of segments used, which were selected with consideration for the existing geological conditions.

Following the construction of each ring, the space between the preliminary support and the rock mass was filled with concrete. Next, after the preliminary support was constructed to the height of 2÷2.1 m, the final concrete lining was executed. In the section under analysis, two types of combined linings were used, namely a panel-concrete and a steel-concrete one, depending on the recognised hydrogeological and engineering-geological conditions, which had an impact on the calculated loads of the lining.

The combined lining was formed by two columns closely cooperating due to their rigid connection. The external column constituting the preliminary ground support was constructed from precast reinforced elements (shaft panels) or steel rings designed to take loads several times greater than the shaft panels and to make it possible to construct the final concrete lining of greater thickness. Due to the presence of sections of watered, very weak cohesive and loose formations at great depths, which translated into very high loads, high-performance concretes were used for the construction of the final linings during the sinking of the 1 Bzie shaft. The application of these concretes required the introduction of specific rigours for industrial production, transportation and installation.

4.1. Preliminary ground support made of precast reinforced concrete

The preliminary panel support in the 1 Bzie shaft was constructed in the form of horizontal rings with the radius of 4.75 m, bound together vertically by means of D26 steel bars installed in the PWS200 seats of the individual segments.

Each ring consisted of 18 precast arch segments of 0.2 m in thickness and 0.7 m in height. The segments formed a reinforced concrete structure made of steel reinforcement cast in class C30/37 concrete.

For technological reasons concerning proper cooperation of the preliminary lining with the rock mass, taking into account the shape of the external surface of the panels, the subsequent rings were installed with horizontal shifting against each other by half the length of the segment. In order to properly bind and ensure, as fast as possible, a proper cooperation of the preliminary support rings with the rock mass, the space between them was filled with concrete through special openings in the lateral sides of the panels, forming a 0.1 m thick levelling layer. Based on the experience gained during the construction of the preliminary ground support from panels in the 1 Bzie shaft, a series of modifications was made in the panels' design, thus facilitating their installation with the use of a shaft manipulator and – technically – achieving better bonding of the reinforced concrete segments.

4.2. Preliminary ground support made of steel segments

The weak, watered formations present in the area of the Bzie 1 shaft at great depths made it necessary to apply a high-strength lining in the sections where the measured loads exceeded the value of 3.8 MPa. In order to avoid using the cast-iron tubing lining, a range of steel rings was designed and manufactured. The rings consisted of welded steel segments of special design and received legal protection – application filed with the Patent Office of the Republic of Poland (RP) – number P.405285. Each ring consisted of 15 "normal" segments and 3 "closing" segments.

Within the range, the segments differed in the thickness of the metal sheets applied, the type of material and the height varying from 0.5 to 1.0 m.

The vertical and horizontal segments were joined to each other by means of M24 screws.

Figure 3. Preliminary ground support made of precast reinforced concrete

To ensure proper cooperation of the individual segments, special filler with adequate character-istics was applied between the frontal surfaces. In order to properly bind and ensure, as fast as pos-sible, a proper cooperation of the preliminary support rings with the rock mass, the space between the steel lining and the rock mass was filled with concrete through the over-flow system construct-ed by means of DN125 pipes in every third segment, forming a 0.1 m thick levelling layer. Taking into account the necessity to conduct controlled water collection from the rock mass in the watered sections at vertical intervals of 30÷40 m between the rings of the preliminary panel or steel lining, a horizontal circumferential drainage of box structure was applied.

The drainage was used to prevent ponding behind the shaft lining.

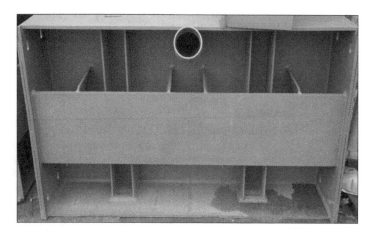

Figure 4. Preliminary ground support made of steel segments

Figure 5. Installation of steel segments in the shaft

4.3. *Final lining made of high-performance concretes*

The final lining that constitutes the internal column was made in the 1 Bzie shaft over the section of deposit overburden formations using class C30/37 to C50/60 concretes. Concrete with the maximum class of C35/45 was used up to the depth of approx. 510 m. However, due to the presence of sectionally watered, very weak cohesive and loose formations below the depth of 510 m, which translated into very high loads, laboratory tests were carried out, followed by test runs with the use of high-performance concretes.

The positive results of the tests were the basis for conditional application of these concretes on the lining of the 1 Bzie shaft. The application of high-performance concretes required the introduction of specific rigours for industrial production, transportation and installation in a shaft. Some of the conditions concerning the production of special high-performance concretes have been presented below:

1) The BBW concretes were produced on the basis of special recipes. Proper amounts of silt fractions were used, along with modern chemicals modifying the rheological properties of the mixture. The correctness of the recipes was confirmed by tests.
2) The constituents were mixed in a planetary mixer at the concrete-mixing facility.
3) The production process consisted of appropriate sequences for the dosage of concrete mixture constituents, and the mixing time could not be shorter than 120 seconds due to the complete activations of additives and chemical admixtures.
4) The concrete mixture had the following parameters:
   - slump flow 650÷700 mm,
   - no segregation of components 30 seconds after stabilisation of the mass,
   - no plastic settlement and bleeding 60 seconds after making test cubes,
   - the permissible consistency drop 60 minutes after mixing up to 5%,
   - apparent density of the mixture 2,200÷2,500 kg/m^3,
   - air content 1.1÷2.5%,
   - temperature of the mixture 18÷26°C,
   - fulfilment of the hydro-technical concrete criterion with regard to concreting in the aquatic environment.

5) The hardened concrete had the following parameters:
   - guaranteed strength class from C45/55 to C50/60,
   - exposure class XD3, XA3,
   - water tightness W12.
6) The time of transportation and unloading of the concrete (transportation to the shaft, quality controls, unloading in the shaft along with placement behind the steel formwork) up to 120 minutes.

CONCLUSIONS

Sinking of the 1 Bzie shaft in difficult geological conditions while maintaining safety of the works required the application of combined special methods, which made it possible to provide access to the rock mass in the shaft shortwall and keep the sidewalls exposed until the lining proper was constructed for the subsequent section of the excavation. Sinking of the shaft in such conditions also required the application of a shaft lining with proper strength, as well as the introduction of suitable modifications into the lining design to guarantee proper cooperation with the rock mass.

REFERENCES

[1] Standard PN-G-05016:1997: "Mine Shafts – Lining – Loads".
[2] Standard PN-G-05015:1997: "Mine Shafts – Lining – Principles of Designing".
[3] Archive materials of KOPEX – Przedsiębiorstwo Budowy Szybów S.A.
[4] Gamaxbeton Archival Materials Concerning the Production of Special High-Performance Concretes and Their Application in a Mineshaft.

International Mining Forum 2015, Kicki et. al. (eds) © 2015 Taylor & Francis Group, London, UK. ISBN 978-1-138-02820-3

Repair of Shaft Brick Lining within Clay-Stone Zone and Proposal of Alteration of the Polish Standard PN-G-14002:1997

Roland Bobek, Tomasz Śledź, Adam Ratajczak
JSW S.A. KWK "Knurów-Szczygłowice"

Wojciech Lekan
LW "Bogdanka" S.A.

Piotr Głuch
Silesian University of Technology, Gliwice

SUMMARY: In unfavourable hydro-geological – mining conditions, shaft brick wall is exposed to a number of disadvantageous influences resulting in its final destruction. In practice, frequent destruction of the brick shaft lining is observed. Implemented solution of the repair of shaft section with use of concrete brick lining has been presented in this work, including proposal of alteration of Polish standard PN-G-14002:1997 entitled: Mining. Concrete bricks used for mine linings, requirements and examinations.

KEYWORDS: Exploitation, lining, shaft lining

1. INTRODUCTION

Since the seventies, shaft linings are usually made in form of monolithic concrete linings. Before the Second Word War brick was used as basal material for shaft lining building, and later concrete bricks were used. Long lasting exploitation of shafts and their exposition onto various influences caused that the linings are often destroyed and break downs in form of mining catastrophes or damages dangerous for exploitation are possible. Repaired shaft was sank in the period 1958–1962 with use of the special rock mass freezing method, and the second shaft in a distance of 100 m was sank in two stages in the period 1908–1915 down to a depth of about 490 m, and in the period 1943–1948 to the depth of 677 m. In long exploitation time, occurrence of numerous damages of the shaft lining, particularly within sections built of brick, were observed.

2. ANALYSIS OF HYDRO-GEOLOGICAL AND MINING CONDITIONS
   IN THE AREA OF SHAFT LINING DAMAGE

General geological structure of the shaft zone at the depth from 150 m to 153 m where the shaft lining was damaged and repair works were needed is shown in Figure 1.

Overburden layers and carboniferous beds occur in the vertical geological stratigraphical-lithological profile of the shaft section. The overburden beds (Quaternary and Tertiary) occur down to the depth of 311,5 m.

Quaternary beds are developed as sands and clays occurring at the depth of about 32,0 m. Sandy layers occur mainly in the lower part of Quaternary beds at the section from 15,65 to 32,0 m, whereas its upper part is built of sandy clays.

Tertiary beds in the shaft region occur at the section from the depth of 32,0 m to the depth of 311,5 m, reaching thickness of about 279,5 m. They consist mainly of clays with intercalations of sands, silts and weakly-cohesive sandstones. Sand and sandstone intercalations of the thickness bigger than 0,5 m have been fund at eight horizons. Sandy intercalations are water saturated. 3,0 m thick layer with gypsum and dusty marls occur at the depth of 304,5 m.

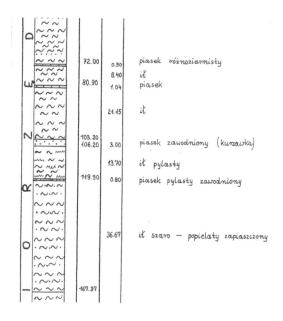

Figure 1. Part of geologic profile of shaft in the region of designed repairing of shaft lining, depth 150–153 m

Carboniferous beds are built of alternate layers of clayey and sandy shales, sandstones and hard coal seams. Sandstones reaching bigger thicknesses prevail in the profile of Carboniferous beds.

In the shaft region, water-bearing layers occur within Quaternary, Tertiary and Carboniferous beds.

Based on the results of hydro-geological examinations executed in the testing bore-hole drilled in May 1953 we can state that all sandy inserts and intercalations occurring within Quaternary and Tertiary beds are water-saturated.

Within Tertiary beds, among clays, occur inserts and intercalations of water-bearing sands and sandstones. Inserts of water-bearing sands of quicksand character having thickness exceeding 0,5 m have were found at the following depths:
- 2,5 m – layer thickness of 0,5 m,
- 81,9 m – layer thickness of 1,0 m,
- 106,2 m – layer thickness of 3,0 m,
- 120,7 m – layer thickness of 0,8 m,
- 226,4 m – layer thickness of 2,1 m.

Depth of the stable water table of individual water-bearing layers is diversified and it gradually grows up with the depth of layer. During hydro-geological examinations, inflow from mentioned layers ranking from 6,1 l/min to about 70 l/min has been proved, with depression reaching usually to the floor of the water-bearing water. At the shaft section in question, no water discharge from behind the shaft lining was proved, and only local moistures were observed. Carboniferous water-bearing stage related with layers of vary-grained and coarse-grained sandstones is in great degree drained via mine workings near the shaft. At the section of Carboniferous layers the shaft is dry.

3. SHAFT MINING CONDITIONS

The shaft in question has diameter of 6,0 m and plays role of outtake-downcast shaft located on the area of mining field "Zachód". Shaft depth amounts for 694,62 m. Inlets for shaft bottoms and chambers (7 pieces) have been made in the shaft.

Actually, since the year 2013, repair works related with change of the shaft function into mate-rial-transport shaft, are conducted. The shaft is located at N-W from the outcrop of hard coal seams of the group 300 and partially of the group 400. In the region of shaft pillar mining exploitation has been conducted in the following seams: 414/3, 415/2, 415/3, 415/4, 416, 418, 419, 501, 502/1, 502/2, 504dw, 507, 510, 613, 615 and 617. The exploitation was conducted with longwall system with roof cut and fill, and rarely with longwall system with hydraulic filling.

4. DAMAGES OF THE SHAFT LINING, GEOTECHNICAL ROCK PARAMETERS
   AND LINING PARAMETERS IN THE AREA OF DAMAGES
   AT THE HORIZON 150 M TO 153 M

During works related with reinforcement disassembly and equipment needed for further recon-struction and shaft function change, occurrence of local exfoliations of brick wall at the horizon be-tween button 51 and button 52, have been proved. The exfoliations occurred practically on whole shaft circumference, and their depth reached to about 20 cm. Part of the destroyed lining (exfoliat-ed bricks) was broken down in result of work conducted in the shaft (Fig. 2).

Figure 2. Damaged brick wall
on the deep of 20 cm

The following rock layers occur in the region of shaft damage (according to bore-hole data):
–   water saturated sand (quick sand) from depth 103,20 m to 106,20 m – thickness of 3 m;
–   dusty clay from depth 106,20 m to 119,90 m – thickness of 13,70 m;

179

- water saturated dusty sand from depth 119,90 m to 120,70 m – thickness of 0,80 m;
- gray-cinereous sandy clay from depth 120,70 m to 167,37 m – thickness of 36,69 m.

According to building log, shaft sinking on the depth of 146 to 170 m, at the depth 50–153 m, shaft sinking rate was relatively low, and time between temporary lining completion and execution of final lining reached about 49 days.

In the region in question, no occurrence of water saturated rock mass in form of water discharge was proved during local inspection. Only local moisture and water discharge from higher horizons related with occurrence of water-bearing horizons, was proved.

For needs of shaft repair design at the section from 150 to 153 m, i.e. in the area of buttons 51 to 52, the basic parameters are as follow:
- γ – bulk density, $\gamma = 0,22$ MN/m^3, assumed as mean,
- $\Phi^{(n)}$ – internal friction angle, $\Phi^{(n)} = 25°$,
- H – depth, for calculated load $H = 153$ m.

Computational load uniform for the lining was assessed at the level = 1,31 MPa.

Samples of exfoliated brick wall were taken, and its strong exfoliation and full water saturation were proved. Compressive tests of the samples taken proved their destruction at the mean compressive strength of 12,1 MPa.

The main cause of developed shaft damages (diameter of 6,0 m) comprise:
- high water saturation of the shaft brick wall,
- low destruction resistance of brick having excessive water absorbability.

Before shaft repair works, prompt temporary protection against shaft lining exfoliation, has been built. Preliminary protection of the shaft lining in fractured zones was designed, and steel mesh was bolted to the lining, with anchors glued in sections 0,5 m long to the steel mesh, including installation of longitudinal channel bars 1,2 m long, bolted to the lining in sections every 1,5 m along shaft circumference, which were bolted to the lining. Scheme of this solution is shown in Figure 3, and its view is shown in Figure 4.

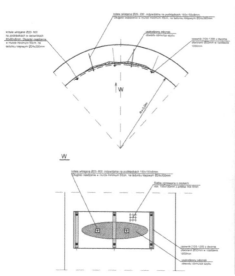

Figure 3. Scheme of initial protection of shaft lining by steel mesh, bolt and channel bar

Figure 4. The view of shaft lining by steel mesh, bolt and channel bar

5. SOLUTION OF THE REPAIR PROCEDURE OF THE DAMAGED SHAFT SECTION

Repair of the damaged shaft fragment on the section 3 m long, at the depth from 150 m to 153 m should assure:
- reconstruction of full lining loading within repaired section,
- leak tightness of the shaft lining analogical the conditions over and below repaired section,
- suitable resistance of the lining material to aggressive environment,
- work safety.

Based on the experience gained in the shaft repair works it was assumed that repair of damaged shaft lining at the horizon from 150 m to 153 m via application of concrete brick lining is fully well-grounded.

The procedure comprises replacement of the existing lining with shaft concrete bricks according to Polish standard PN-G-14002:1997 – Mining. Concrete bricks used for mine excavation linings. Requirements and examinations. The concrete bricks should be laid with use of cement mortar of high resistance to aggressive environment and long durability. Concrete bricks should be joined with use of cement mortar of viscous consistence.

Concrete bricks should be selected from Table 3 of the Polish standard PN-G-14002:1997, and strength should be selected according to Polish standard PN-G-05015:1997 (see Tab. 4).

Characteristic compressive strength of the wall made of concrete bricks should be selected from Polish standard PN-G-05015:1997 (Tab. 1).

Table 1. Characteristic compression strength of wall made with concrete bricks by PN-G-05015:1997

Class	Compressive strength of mortar (Φ 0,8 cm) in MPa (mortar brand)		
	8 (M12)	10 (M15)	12 (M20)
200	8,9	9,2	9,4
250	10,9	11,2	11,4
300	12,8	13,1	13,4
350	14,7	15,0	15,4

Required thickness of the shaft lining within repaired section is calculated according to Polish standard PN-G-05015:1997 – Mining Shafts. Lining. Design rules, from the formula:

$$d = a \cdot \left(\sqrt{\frac{k}{k - p \cdot \sqrt{3}}} - 1 \right) \tag{1}$$

where: a = 3 m – inner shaft radius, p = 1,31 MPa – uniform computational loading acting onto shaft lining, $k = R_{nb}/s$ – permissible strength of concrete brick wall in MPa, R_{nb} – characteristic compressive strength of the concrete brick wall, according to Table 1, for individual mortar brands, s – factor of safety.

According to Polish standard PN-G-05015:1997, factor of safety should amount for 2,5. In case of favourable hydro-geological conditions, use of this factor value s = 2, is permissible.
Calculated (required) values of the shaft concrete brick lining for various classes with safety factor s = 2 and mortar of the brand M20, are shown in Table 2.

181

Number of rings of the concrete brick lining of type BS2 for shaft diameter of 6,0 m, according to Table 4 (Polish standard PN-G-14002:1997) at their length of 36 cm can be calculated from formula:

$$n_p = d/36 \qquad\qquad\qquad (2)$$

Calculated and required shaft lining thicknesses used for repair of the shaft section (BS2) are shown in Table 2.

Table 2. The required thickness of shaft lining in repaired shaft with wall made of concrete brick BS2

Lk	a Shaft radius [m]	Class of concrete brick Acc. PN-G- -05015:1997	d Calculated thickness of the lining [m]	n_p Number of rings	Thickness in concrete bricks	d_{bet} [cm]
1.	3	200	1,169	3,25	3,5	36+1+36+1+36+ +1+18 = 129 cm
2.	3	250	0,865	2,4	2,5	36+1+36+ +1+18 = 92 cm
3.	3	300	0,688	1,89	2	36+1+36 = 73 cm
4.	3	350	0,571	1,586	1,5 brick	36+3+18 = 57 cm

Shaft concrete bricks of class BS2 should be used for repair of the brick wall:
– minimum 300, i.e. B30, what corresponds to concrete class C25/30,
– or class 350, i.e. B35 (actual B37), what corresponds to concrete class C30/37,
where:
– C25/30 and C30/37 – classes of concrete according to Euro-code 2,
– B30 and B35 classes of concrete according Polish standard PN-B-03264-1999 – concrete, reinforced concrete and pre-stressed constructions. Static calculations and design.

Application of concrete bricks of the class minimum 300 or 350 with mortar brand M20 was designed.

Application of two computational variants of the execution of brick wall repair via re-bricking is possible:
– variant I – re-bricking of the brick wall with double thickness of concrete brick BS2 of thickness 73 cm, class minimum 300 (B30 – C25/30) with mortar type M20,
– variant II – re-bricking of the brick wall to thickness of 1,5 of concrete brick with thickened vertical joint of two concrete bricks BS2 type C and BS2 type PP (half) of the thickness 57 cm class minimum 350 (B37 – C30/37) with mortar class M20.

In the second case, concrete bricks of the class 350 made of concrete C30/37 have higher concrete class than mentioned in Polish standard PN-G-05015:1997, what means that real concrete brick should amount for 370, what with application of mortar M20 improves compressive strength of the wall up to value of:

$$R_{mb} = 0,04x + 1,4 = 0,04 \cdot 370 + 1,4 = 16,2 \text{ MPa}$$

Permissible strength of the concrete brick wall built of concrete B37 (C30/37) obtained from approximation of the data from Table 4, according to Polish standard PN-G-05015 amounts for

1997, for $s = 2$ permissible stress of the wall built of concrete brick of the class B37 (C30/37) amounts for: $k = 16,2 / 2 = 8,1$ MPa.

Computational thickness of the concrete brick lining of class B37 according to $p = 1,31$ MPa amounts for:

$$d = a \cdot \left(\sqrt{\frac{k}{k - p \cdot \sqrt{3}}} - 1 \right) = 3 \cdot \left(\sqrt{\frac{8,1}{8,1 - 1,31 \cdot \sqrt{3}}} - 1 \right) = 0,535 \, \text{m} \tag{3}$$

Finally, in case of application of concrete bricks made of class C30/37 concrete corresponding to class B37, and according to Polish standard PN-G-05015:1997corresponding to class 370, with use of cement mortar of brand M20, computational thickness of the reconstructed brick wall amounts for $d_{min} = 53,5$ cm.

Number of concrete bricks 36 cm wide amounts for $n_b = 53,5 / 36 = 1,486$ pieces. Thus thickness of the brick wall made of 1,5 concrete brick having dimensions $d_b = 36 + 1 + 18 = 55$ cm is satisfactory,

where:
- 36 cm – thickness of the inner ring (from the shaft side) made of concrete bricks of class C30/37(B37), – concrete bricks BS2 of type C, according to Polish standard PN-G-14002:1997,
- joint between concrete brick rings of the thickness from 1 cm to 1,5 cm,
- 18 cm – thickness of the outer ring of concrete brick having parameters of the C30/37(B37) class concrete, – concrete bricks BS2 type PP according to Polish standard PN-G-14002:1997 [1].

Proposed final solution of the shaft lining is shown in Figure 5 and the solution in realization phase in Figure 6.

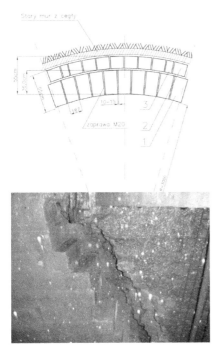

Figure 5. Wall made of concrete brick BS2, second variant, thickness 55 cm, concrete class – C30/37, mortar – M20.
1 – betonit BS2-C-20,7×18,7×14-36,
2 – betonit BS2-PP-20,7×19,7×14-18,
3 – betonit BS2-PP-½

Figure 6. The view of bricklaying by use concrete brick

6. PROPOSAL OF ALTERATION OF THE POLISH STANDARD PN-G-14002:1997

Shaft lining made of bricks according to Polish standard PN-G-05015:1997 [2] should be calculated with use of load capacity limiting states, thus on the basis of Polish standard PN-B-03002::2007 [3], characteristic strength f_k is determined. According to Polish standard PN-B-03002:2007 characteristic strength of brick walls is dependent on:
– group of products depending on strength parameters (from 1 to 4),
– categories of elements depend on quality assurance control (cat. I and II),
– mortar brand, accepting Brand (special) M_d, where: d compressive strength bigger than 25 MPa.

In dependence of the production quality controls, brick wall elements are classified into category I or II:
– category I comprises brick elements, which according to manufacturer's declaration possess suitable compressive strength – and factory's quality control assures that probability of mean compressive strength lower than declared is no higher than 5%;
– category II comprises brick elements, which according to manufacturer's declaration possess medium compressive strength and the other requirements of the category I are not satisfied.

Normalized compressive strength f_b is considered as basal parameter characterizing the brick elements in question.

Elements of brick constructions are dimensioned by the method of limited states. Three partial safety factors were used in the calculations: γ_f – referred to loads, γ_m – referred to brick wall properties, and γ_s – referred to steel. Value of the computable strength of the brick wall is determined as the ratio of brick wall characteristic strength and partial safety factor γ_m.

Value of this factor is determined in dependence on quality control category of wall elements, type of the used mortar and category of the conducted works.

The following categories are distinguished:
– category A of conducted works, when brick works are conducted by suitable trained personnel under supervision of brick foreman, factory manufactured mortars are used, and in case when mortars are prepared in situ, dozing of components is suitably controlled, and mortar strength and quality of works are verified by the inspector of investor's supervision.
– category B of conducted works when conditions of category A are not satisfied; In such case supervision of working quality can be executed by suitably trained person, authorized by the contractor. Values of the brick wall safety factors γ_m, used in calculations of the construction in temporary situations, are gathered in Table 3.

In unique situations, independently on brick wall category and category of executed works, the following factors can be assumed: with reference to the brick wall $\gamma_m = 1,3$, with reference to anchorage of reinforced steel 1,15, with reference concrete steel $\gamma_m = 1,0$.

According to standard of the shaft lining calculation PN-G-05015:1997 for concrete brick walls, the compressive strength depends on hydro-geological conditions, and value of safety factor is assumed as $s = 2,0$ or $s = 2,5$.

Mining experience indicates that shaft linings made of concrete brick are more durable than lining made of bricks. Advance rate of concrete production in scope of obtaining high strength classes and required durability in dependence of the environment aggressiveness, for concrete [3], allows production of high class concrete bricks. Proposed adaptation and widening of the class of produced concrete bricks are shown in Table 4.

Table 3. The values of partial factory of wall safety according to PN-B-03002:2007 [4]

Material	Working categories	
	A	B
Walls made of concrete bricks of category I and designed mortar (strength experimentally controlled)	1,7	2,0
Walls made of concrete brick elements of category I and prescribed mortar (strength determined on the basis of volumetric proportion of components)	2,0	2,2
Walls made of concrete brick elements of category II and optional mortar type	2,2	2,5
Anchorages of concrete steel	2,0	2,2
Concrete steel	1,15	
Headers prefabricated according to PN-EN 845-2	1,7	

Table 4. Characteristic strength of wall made of concrete brick with corresponding concrete class C16/20 to C40/50

Concrete class	Concrete brick class f_b [MPa]	Mortar brand			
		M12	M15	M20	M25
C16/20, B20	20	8,9	9,2	9,4	-
C20/25, B25	25	10,9	11,2	11,4	-
C25/30, B30	30	12,8	13,1	13,4	13,8
C30/37, B37	37	15,49	15,79	16,2	16,6
C35/45, B45	45	18,58	18,88	19,4	19,9
C40/50, B50	50	20,51	20,81	21,4	22,0

Possibility of neglecting determination of lining thickness with accuracy of half of the concrete brick width is important to assure proper arrangement of the concrete bricks in the wall. Favourable operation of the concrete brick wall is obtained in case of single-layer walls. In practice, wedge-shaped of the square dimensions 14 cm×14 cm from the shaft side (with permissible mass of 30 kg) can be up to 58 cm long (for shaft of diameter of 6,0 m – as in case of repaired shaft lining).

With respect to technology development via application of the category I of the production of the wall elements and works (brick wall) of category A, the safety factor can be reduced with 15% (analogically to possibility of 20% reduction of the material coefficient γ_m for brick wall) in favourable hydro-geological conditions.

For the project of the shaft repair in question, for concrete brick 37 with mortar M20 and in good hydro-geological conditions (lack of water outflows from behind the wall), for $s = 2,0 - 2 \cdot \cdot 0,15 = 1,7$ we obtain permissible compressive strength $k = 16,2 / 1,7 = 9,53$ MPa.

Required thickness of the concrete brick lining amounts for:

$$d = a \cdot \left(\sqrt{\frac{k}{k - p \cdot \sqrt{3}}} - 1 \right) = 3 \cdot \left(\sqrt{\frac{9,53}{9,53 - 1,31 \cdot \sqrt{3}}} - 1 \right) = 0,436 \, m \tag{4}$$

Sufficient thickness of the concrete brick wall (gradation of the concrete bricks length every 5 cm) amounts for 45 cm – mass of about 23 kg, class 37 with mortar M20.

Introduction of alterations due to actual technological state and commonly used standards allows reduction of the repaired brick wall with 10 cm.

FINAL CONCLUSIONS

For needs of the repair of shaft wall from horizon 150 to horizon 153, the solution of the lining reconstruction and obtaining its strength parameters allowing transfer of the acting load, i.e. safe shaft operations, has been proposed.

The selection of the solution should take into consideration:
- technical potential,
- technological potential,
- economical means,
- work safety,
- rep air working time,
- existing experience.

In the case in question, the following has been designed and applied – variant of the replacement of brick lining into concrete brick of thickness equal to 1,5 of concrete bricks of the type BS2 class 370 (concrete C30/37, B37) – wall thickness about 55 cm (Fig. 5).

The solution allows full reconstruction of the shaft lining strength and needed load capacity assuring lining stability.

On the basis of already existing standards, practical experiments and technological means, the alteration of Polish standard PN-G-05015:1997 in scope of classes and dimensions of concrete bricks, mortar brands and obtained strength, with detailed reconnaissance of hydro-geological conditions, and calculated safety factor s on the level 1,7, allow efficient designing of the shaft lining thickness, particularly in its reconstruction. Taking into consideration necessity of the repair with replacement of numerous fragments of damaged shaft linings, the authors suggest that alterations of the Polish standard PN-G-05015:1997 is necessary.

REFERENCES

[1] PN-G-14002:1997: Mining. Concrete Bricks or Mining Road Linings. Requirements and Examinations.
[2] PN-G-05015:1997: Mine Shafts. Lining. Design Procedures.
[3] PN-B-03264:2002: Concrete. Reinforced Concrete and Pre-Stressed Constructions. Static Calculations and Design Procedures.
[4] PN-B-03002:2007: Brick Construction and Design Procedures.

Repair of Brick Shaft Lining in the Zone of Mudstone with Proposition of Changes in PN-G-14002:1997

Repair of shaft lining in layers of mudstone with watered layers of sands has been carry out by utilization of high strength wall made of concrete brick. For this construction has been used concrete brick of concrete class C30/37 and cement mortar M20.

The use of concrete brick with increased strength has allowed design the wall 55 cm thick, with does not require interruption of existing lining. The changes of standards have been proposed, on the base of experiences and analysis of the existing standards. These changes allow use of concrete brick made of higher class of concrete and individual specification of brick length every 5 cm. For design purpose has been proposed reduction of safety factors s by its decreasing about 15% (up to value $s = 1,7$) at first category of wall element and category A of work performance.

International Mining Forum 2015, Kicki et. al. (eds) © 2015 Taylor & Francis Group, London, UK. ISBN 978-1-138-02820-3

Application of Steel Segments in the Reinforcement of Shaft Lining at a Section of Sandwater Area

Aleksander Wardas
JSW S.A. KWK "Knurów-Szczygłowice"

Wojciech Lekan
LW "Bogdanka" S.A.

Piotr Głuch
Silesian University of Technology, Gliwice

SUMMARY: Installation of steel segments is one of the methods for the reinforcement of the shaft lining. Using adjusted steel segments, it is possible to make a lining that will be characterized by the parameters of the final lining. The paper presents the structural solution and the installation method of steel segments reinforcing the shaft lining at a section of sandwater area with existing heavily damaged brick lining. After ripping out the damaged elements, a galvanized steel shell comprising of segments and bonded with the existing lining has been bolted to the wall using oblique bolts driven to a depth of ~25 cm.

KEYWORDS: Exploitation, lining, shaft lining

1. INTRODUCTION

Difficult geological and mining conditions in case of driving and maintaining vertical headings, such as shafts, ore passes and storage reservoirs result mostly from large depths of these headings. The behaviour of the lining is significantly affected by the dimensions of the headings reaching up to 12 m in cross-section.

Further factors that are decisive in the use of lining are:
- low compressive strength of the rocks in the 20÷30 MPa range;
- occurrence of faulting, folding, overthrusts, dipping strata and other displacements causing considerable degradation of rock structures;
- possible sliding at the points where the strata meet;
- occurrence of swelling rocks around the shaft heading, such as: silts, clays or marls increasing their volume e.g. in freezing.

Difficult geological and mining conditions should also include instances of strong action of aggressive environment on the shaft lining, which decreases its strength and causes damage to its structure.

Problems related to the maintenance of the stability of shaft linings, ore passes and storage reservoirs should be viewed with consideration to the impact of exploitation and mining works conducted in the vicinity of these headings.

In the mining practice, the damages of the shaft linings are repaired by i.e.:
- the reinforcement of the lining with steel rings,
- the reinforcement using bolts with meshes and rings,
- shotcreting after removing the damage from the lining,
- exchanging the lining with a new structure,
- installation of a steel shell at the entire section of the shaft along with bolting it to the wall or the rock mass.

Steel rings used in the reinforcement of the shaft may be both double-shell – comprising of an internal and external ring – and single-shell structures [2], [3]. In both cases, the structures of steel shells allow to bolt them to the existing lining or to the rock mass.

2. GENERAL ANALYSIS OF THE HYDOGEOLOGICAL AND MINING CONDITIONS

Geological conditions

The rock mass of the reinforced shaft in the area of the bunton No. 10 is comprised of Quaternary strata, represented by:
- from the depth of 13.0 m do the depth of 22.5 m – hydrated plastic silt with high content of quartz dust and fine-grained sand,
- from the depth of 22.5 m do the depth of 25.7 m – water-bearing sand (*sandwater*),
- from the depth of 25.7 m do the depth of 26.0 m – water-bearing silty clay loam (siCl).

In the area of the bunton No. 10, that is 23÷25 m, no tectonic displacements have been indicated. The general geological structure in the depth range up to 50 m in the area of the bunton No. 10, has been presented in Figure 2.1.

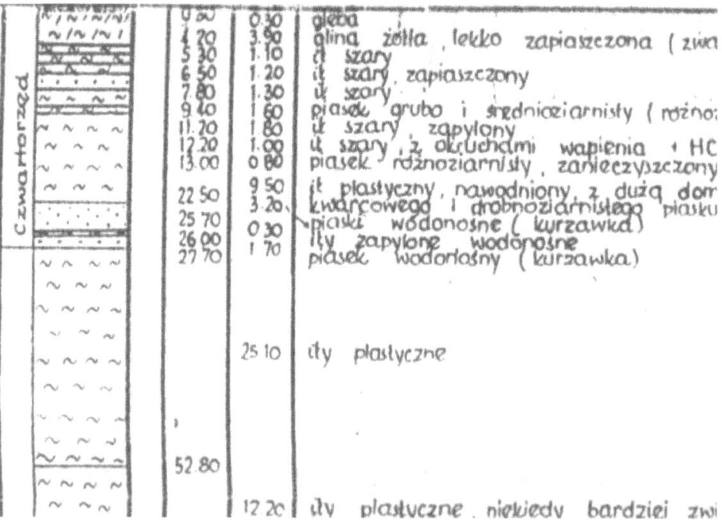

Figure 2.1. The overall geological cross-section through the layers
of the Quaternary in the shaft to depth of about 55 m

188

Hydrogeological conditions

In the Quaternary formations, that is, from the depth of 7.8 m to the depth of 27.7 m, three water-bearing horizons have been separated:

I water-bearing stratum
Is located from the depth of 7.8 m up to the depth of 9.4 m and is constituted by coarse and medium grained sandy clay clSa – the shaft lining from the opening of the shaft up to 15 m is dry – the level has been drainaged.

II water-bearing stratum
Is located at the depth from 12.2 m to the depth of 13.0 m and is constituted by variously-grained clSa. At the stage of shaft sinking, the water inflow from that section was ~ 100 l/min. Currently at this section the shaft is dry and has probably been also drainaged.

III water-bearing stratum
Is located at the depth from 22.5 m to the depth of 27,7 m and is constituted by silty sand (siSa) and fine-grained clSa. In the process of sinking, the first inflow of water occurred at the depth of 20.5 m and sandwater entered the shaft at the depth of 22.5 m. Currently, rapid inflows of water at a rate of 10 l/min are observed, which occur at depths between 9.0 and 27.0 m and are most intense on the eastern side of the shaft. The water flows down on the wall lining to a trough installed at the depth of 81 m, and then is carried to the level 450, where the inflow is measured. In 2010, the inflow ranged from 3 to 7 l/min.

The analysis of the water samples acquired from the trough at the depth of 81 m, has indicated that the water is weakly mineralized, hard with a total hardness of 9.86 meq/dm^3 and prevalence of carbonate harness, weakly basic. The performed analysis of the water has shown that the water is of calcium-sulfate type. In line with the PN-EN 206-1:2003 standard, such water is denoted by XA1 exposure class characterizing weakly aggressive environment in relation to concrete.

3. CALCULATION OF THE LOAD OF THE SANDWATER STRATUM ON THE SHAFT LINING

The formulas for the calculation of load on the VI shaft have been specified mostly using the "PN-G-05016:1997. Szyby górnicze. Obudowa. Obciążenia". Standard.

In case of rocks occurring in the form of sandwater, the following formula is admissible in the calculation of load:

$$p_s = \gamma_{nk}^{(n)} \cdot H \tag{3.1}$$

where: $\gamma_{nk}^{(n)} = 0.0127$ to 0.0135 MN/m^3.

Giving consideration to the presence of the sandwater stratum, at the depth of $H = 27.7$ and assuming the specific gravity of sandwater $\gamma_{nk}^{(n)} = 13.5$ kN/m^3 the pressure on the lining amounts to:

$$p_s = 13,5 \cdot 27,7 = 373,95 \text{ kN/m}^2.$$

In the calculations p_s – the calculated pressure on the lining has been assumed to be $p_s = 0.4$ MPa.

The bearing capacity of the existent brickwall lining characterized by the calculated strength of the wall:

$R_m = 2.79$ MPa.

Amounts to:

– in case of the wall of $d = 0.77$ m in thickness:

$p_{ob} = R_m \, d \cdot l/r_z = 2.79 \cdot 0.77/4.37 = 0.491$ MPa;

– in case of the wall of $d = 0.51$ m in thickness:

$p_{ob} = R_m \, d \cdot l/r_z = 2.79 \cdot 0.51/4.11 = 0.346$ MPa.

Considering the fact that the exact thickness of the lining in the area of the bunton No. 10 is not known, due to difficulties in the sinking of the shaft at the depth of ~25 m and due to the occurring deformations and the level of damages in the lining, namely scaling of bricks and washing out of joints, the bearing capacity of the shaft in its current state should be increased.

4. THE CONCEPT OF SHAFT REPAIR AT THE 19.1 ÷ 29.16 M SECTION

The structural solution and the method used for the reinforcement of the shaft lining, and in fact, its repair and ensuring its strength (bearing capacity) against horizontal loads, consists in the application of steel segments made of 10 HAV plate (or similar) in line with PN-H-84017:1983 with a strength of $f_d = 290$ MPa. The segments should be completely galvanized to ensure at least 30 years of exploitation.

As an example, it is admissible to use S355J2 steel type, the parameters of which should be:
– $R_e = 340$ MPa, – $R_m = 520$ MPa, $-A_5 = 30\%$, – $KV_2 = 163$ J (in the temperature of -20°C).

Steel segments made of 20 mm steel are made into steel rings. 12 segments are used in each of the steel rings. The height of the segment and the ring is 300 mm. The segments are bolted with screws (M 27 bolts at joint). The rings, in turn, are connected with 36 screws (three screws per segment). The vertical joints of the segments and the distribution of screws has been designed so as the segments in each of the subsequent rings may be moved by at least 10°. The solution utilizes a structure with segments welded from a 20 mm sheet with a shape of a channel bar section. These rings have a diameter of $D_{zw} = 7180$ mm and are bonded to the shaft lining with a bond applied from top after the assembly.

The bond, which is injected behind the ring, tightly fills the space between the ring and the existent lining. The bond should possess the following qualities:
– high penetration of fractures and fissures in the lining, and thus reinforcing its structure,
– resistance to washing out by water flowing down on the lining of the shaft,
– high adhesiveness to steel and the existing lining material, as well as high leak tightness.
    It is advised to apply:
- cement binders (micro cements),
- silicone resins with a strength of approximately 30 MPa.
    The design assumes that each element shall be bolted with at least 2 oblique bolts with a length of at least 450 mm at the thickness of the lining that is approximately 250 mm.
    The joints of the segments made with three 8.8 class M27 bolts should be displaced in relation to one another by at least 1/3 of the segment's perimeter.

The first of the rings is founded on a base ring at the depth of 29.16 m – this depth should be indicated by mine surveyors and levelled correctly.

The exact level determined at the stage of preparing the base ring, should give consideration to the technology of works in the shaft and the distribution of buntons. The horizontally installed base ring should not be founded above the 29.16 level. The base ring is secured with bonded steel anchors fixed to the betonite shaft lining. Considering the fact that the thickness of the betonite lining is 74 cm, the foundation of the ring has been designed as mounted with adhesive bonded anchors with 22 mm diameter and a length of 450 mm, installed perpendicularly to the ring. The base ring should be levelled so as to ensure the correct placement of the reinforcement rings in the shaft. The structural solution of the reinforcement has been presented in the layout drawings of segments and reinforcement rings.

Figure 4.1. Steel segment with a height of 300 mm with a 20 mm thick sheet, view from inside the shaft

Figure 4.2. Steel segment with two bearing bolts, view from outside the shaft

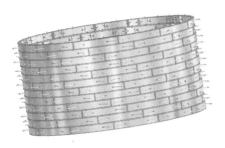

Figure 4.3. A section of rings used in the reinforcement of the shaft lining

Figure 4.4. Rings with bolts arranged with displacements of vertical joints by 10°

191

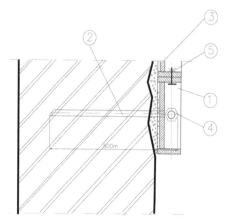

Figure 4.5. Cross section of the base ring bolted to the betonite lining at the 29.16 m level (below the sandwater level):
1 – base ring, 2 – bolt,
3 – reinforcement ring,
4 – horizontal screw,
5 – vertical screw

5. STRUCTURE AND CALCULATIONS FOR THE BEARING CAPACITY AND DISTRIBUTION OF THE REINFORCEMENT RINGS

The ring's bearing capacity is calculated with consideration given to:
- strength,
- stability,
- local stability losses,
- non-uniform loads.

The strength of the lining made of steel rings is specified using the following formula:

$$p = \frac{A_c \cdot n_p \cdot f_d \cdot \varphi}{1 \cdot r} \tag{5.1}$$

where: p – external pressure, calculated load, r – radius of the shaft (external radius of the ring), n_p – no. of rings per 1 m of a shaft, A_c – area of a cross-section of a single ring, f_d – calculated strength of the steel of the ring, φ – buckling factor.

The stability of the entire ring is established using a formula which gives consideration to the foundation of the ring with the use of a bond and bolts and utilizing the O. Domek's formula.

$$p_{kr} = 2 \frac{\sqrt{K^{(\sigma)} \cdot E \cdot I}}{r} \tag{5.2}$$

where: E – Young's modulus for steel, I – moment of inertia for a cross-section of a ring at 1 m of a lining, r – external radius of rings, $K^{(\sigma)}$ – passive pressure coefficient expressed as MN/m³.

The coefficient has been determined using Gerelkin's formula.

$$K^{(\sigma)} = \frac{E_{mat}}{r_z \cdot (1+v)} \tag{5.3}$$

192

Critical pressure at which loss of stability occurs is additionally calculated using a general formula as provided by BRESS [1]:

$$p_{kr} = \frac{3 \cdot E \cdot I}{R^3(1-v^2)} \tag{5.4}$$

where: E – Young's modulus for steel, I – moment of inertia for a cross-section of a ring at 1 m of a lining, assuming the installation of 3 rings at a section of 1 m of shaft, R – external radius of rings.

The local loss of stability is specified using the following formula:

$$\sigma_{kr} = k \cdot \frac{\pi^2 \cdot D}{b^2 \cdot h} \tag{5.5}$$

where: b = width of a ring, h = thickness of the plate, k – coefficient depending on the ratio of a and b.

Stresses in the ring for the maximal pressure of p = 0.4 MPa

$$\sigma = \frac{p \cdot r \cdot d}{A_c \cdot \varphi} \tag{5.6}$$

should fulfil the following condition:

$$\sigma_{kr} \langle \sigma \tag{5.7}$$

In the conversion of the stability of the steel ring into the local stability of a one-hinged arch, when the loads are transferred by two segments, the critical radial pressures q_{kr} are specified by the formula (Timoshenko S.P., Gere J.M.: Elastic Stability Theory. Arkady, 1963).

$$q_{kr} = \gamma_1 \cdot \frac{EJ}{r^3} \tag{5.8}$$

where: γ_1 – coefficient acquired from the table, E – Young's modulus for steel, J = moment of inertia of the cross-section, r – internal diameter.

The bearing capacity of screws joining the rings has been verified for the condition of transferring the load specified from the strength of the ring as a three-hinged arch formed by a segment of the ring.

Assuming that an adverse load acts on a single segment, the screws should not be truncated. The force acting on a segment

$$P_{seg} = p \cdot b \cdot L \tag{5.9}$$

where: b – width of a segment, L – length of a segment.

The bearing capacity of an 8.8 class M27 bolt in the wall is:

$$N_{\acute{s}\acute{c}} = 0.58 \cdot N_{zr} \qquad (5.10)$$

The number of screws fastening the segments at joints into the ring is:

$$n_s = P_{seg}/N_{\acute{s}\acute{c}} \qquad (5.11)$$

The geometric parameters of the segment used in the structure of the ring have been presented in Figure 5.1.

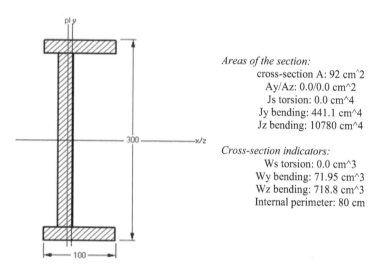

Areas of the section:
cross-section A: 92 cm^2
Ay/Az: 0.0/0.0 cm^2
Js torsion: 0.0 cm^4
Jy bending: 441.1 cm^4
Jz bending: 10780 cm^4

Cross-section indicators:
Ws torsion: 0.0 cm^3
Wy bending: 71.95 cm^3
Wz bending: 718.8 cm^3
Internal perimeter: 80 cm

Figure 5.1. Geometric parameters of segment section

6. OUTLINE OF THE TECHNOLOGY

To conduct the repair of the shaft, a detailed technology of works has been developed.

The construction of reinforcement made of steel rings was first started as a base ring bolted to the betonite lining of the shaft.

The rings have been fixed to the lining of the shaft using bolts as well as using a bonding material applied behind the segments. The base segments bolted to the betonite lining of the shaft, after the installation and levelling, constitute the base ring. The base segments have been bolted to the shaft lining using ø22–450 (M22) bolts perpendicularly to the shaft wall. The rings were bonded to the existent lining using a mineral bonding agent.

The reinforcing rings made of segments were installed after the installation and the levelling of the base ring made of segments. In subsequent rings, the segments were circumferentially displaced by 10° each, so as the joints were not in a single line. The segments of reinforcement rings were bolted using ø22–450 (M22) bolts oblique in relation to the shaft wall, so as the wall does not rupture and allow the contact with the sandwater, while the brickwall is additionally reinforced at the entire perimeter of the shaft.

The rings were bonded to the existent lining using a mineral bonding agent. The bolting was conducted following the bonding of the bonding agent with the reinforcement rings.

Orifices in segments used in installation were cut in locations of the buntons and supports, according to measurements made during the installation.

At single segments located by large water inflows and places where the lining was severely damaged, omissions of oblique bolts were allowed to enable controlled inflow of water from behind the shaft lining.

No sealant injections were performed so as not to allow the accumulation of water behind the lining and any additional load applied on the lining in the higher strata.

Figure 6.1. The view of zinc coated steel sections

Figure 6.2. The view of steel segments installed in the shaft

Figure 6.3. Protected section of the shaft in the sandwater zone with visible surface damage of brick lining

Figure 6.4. Completion of the reinforced section in the steel jacket using a bevel made of adhesive binder

FINAL CONCLUSIONS

1. The repair of the shaft section located in the sandwater area by reinforcement has been conducted based on:
– the analysis of geological and mining conditions in the shaft,
– site inspection and observation of the existing damages,
– determination of the causes of the existing damages,
– the repair plan of the shaft,
– bill of quantities, repair technology outline and the estimated cost of the investment.
2. The reinforcement of the shaft up to the level of 29.16 m up to the depth of 19.1 m has been designed with the application of an independent lining structure, namely steel reinforcement rings installed at the perimeter of the shaft, characterized by increased strength parameters and made of steel segments of galvanized 10 HAV-type plate – in line with the PN-H-84017:1985 standard – with f_d = 290MPa strength or equivalent galvanized.
3. The segments of the ring are bolted using M27 screws and are bonded to the shaft lining using both bolts and by injection of a bonding material which allows for:
– the penetration of fractures and fissures in the lining and reinforcement of its structure,
– protection against washing out with the water flowing down the lining of the shaft,
– the resistance against chemical corrosion,
– high adhesion to steel and concrete,
– high tightness,
– compressive strength of at least 50 MPa.
In practice, a modified cement binder made of mineral materials was applied. Shall this be needed, the constructed reinforcement structure allows for prolonging – both upwards and downwards.
4. The shaft lining repair works were performed under the author's supervision and the design was verified on a current basis which allowed for adjustments to the current geological and mining conditions.

REFERENCES

[1] Głuch P., Szczepaniak Z. 1988: Głębienie szybów. Skrypt Politechniki Śląskiej, nr 1365, Gliwice.
[2] Chmielewski J., Głuch P., Lekan W., Sądej W. 2011: Utrzymanie obudowy zbiorników retencyjnych na dużych głębokościach w warunkach LW Bogdanka. School of Underground Mining, Kraków.
[3] Głuch P., Lekan W. 2013: Nowe konstrukcje stalowych segmentów obudowy szybów, szybików i zbiorników retencyjnych w trudnych warunkach geologiczno-górniczych. School of Underground Mining, Kraków.

Application of Steel Segments in the Reinforcement of the Shaft Lining at a Section of Sandwater Area

Additional reinforcement by means of rings made of zinc coated steel segments has been used for the existing shaft lining made of brick wall in the zone of sandwater on the low depth (approximately 20 m).

After fixing together for the better stabilization of rings, the steel segments have been bonded to the existing shaft lining (after being cleaned) with the use of binding agent and bolted using oblique bolts driven to a depth of 25 cm.

Practical realization of the reinforcement was carried out on non-working days and no problems arising from the applied technology were encountered.

International Mining Forum 2015, Kicki et. al. (eds) © 2015 Taylor & Francis Group, London, UK. ISBN 978-1-138-02820-3

Author Index